Lecture Notes in Networks and Systems

Volume 143

AF823404

The series "Lecture Notes in Networks and Systems" publishes the latest developments in Networks and Systems—quickly, informally and with high quality. Original research reported in proceedings and post-proceedings represents the core of LNNS.

Volumes published in LNNS embrace all aspects and subfields of, as well as new challenges in, Networks and Systems.

The series contains proceedings and edited volumes in systems and networks, spanning the areas of Cyber-Physical Systems, Autonomous Systems, Sensor Networks, Control Systems, Energy Systems, Automotive Systems, Biological Systems, Vehicular Networking and Connected Vehicles, Aerospace Systems, Automation, Manufacturing, Smart Grids, Nonlinear Systems, Power Systems, Robotics, Social Systems, Economic Systems and other. Of particular value to both the contributors and the readership are the short publication timeframe and the world-wide distribution and exposure which enable both a wide and rapid dissemination of research output.

The series covers the theory, applications, and perspectives on the state of the art and future developments relevant to systems and networks, decision making, control, complex processes and related areas, as embedded in the fields of interdisciplinary and applied sciences, engineering, computer science, physics, economics, social, and life sciences, as well as the paradigms and methodologies behind them.

**** Indexing: The books of this series are submitted to ISI Proceedings, SCOPUS, Google Scholar and Springerlink ****

More information about this series at http://www.springer.com/series/15179

Gheorghe Ion Gheorghe
Editor

Proceedings of the International Conference of Mechatronics and Cyber-MixMechatronics - 2020

Springer

Editor
Gheorghe Ion Gheorghe
Mechatronics and Measurement Technique
National Institute for Research
and Development
Bucharest, Romania

ISSN 2367-3370 ISSN 2367-3389 (electronic)
Lecture Notes in Networks and Systems
ISBN 978-3-030-53972-6 ISBN 978-3-030-53973-3 (eBook)
https://doi.org/10.1007/978-3-030-53973-3

This Springer imprint is published by the registered company Springer Nature Switzerland AG
The registered company address is: Gewerbestrasse 11, 6330 Cham, Switzerland

Preface

This book is a collection of papers submitted and accepted for presentation on occasion of the 4th International Conference of Mechatronics and Cyber-MixMecatronics/ICOMECYME held in Bucharest, Romania, during September 10–11, 2020. This conference is envisioned as a forum and an opportunity to researchers, engineers, professors, PhD students, graduate students as well as business representatives from all over the world to present their research results and development activities.

The wide background of the authors of the chapters of the book grants a rich palette of information for the interested parties all over the world, while the careful supervision of the Editor, who is a highly appreciated professor and who is a pioneer of the new science of cyber-mix-mechatronics, will grant the audience access to a world of knowledge that would increase their scientific potential and fuel it to generate, in turn, new ideas from their side.

This book covers a vast range of topics from integrated mechatronics, integronics and adaptronics; cyber-mechatronic and cyber-mixmechatronics; claytronics and cyber-claytronics to smart bio-medical and bio-mechatronic systems; MEMS and NEMS; instrumentation and measurement; smart environmental systems; process monitoring; new materials; technology transfer of high-tech mechatronic in industry; water network monitoring; nano-chemistry, physical chemistry of biological systems; chemical reactions: mechanisms, dynamics, kinetics and catalytic reactions; risk integrated management; gas and oil networks monitoring; high productivity and high added-value industrial services; and other specialized smart fields.

Special thanks to Mrs. Varsha Prabakaran and to Mrs. Holger Schaepe, for making the publishing process a smooth and successful one.

Gheorghe Gheorghe
Conference Chairman

Contents

The Influence of the Rotating Piston Shape on the Sealing Between Two Profiled Rotors of a Rotating Machine that Transports Fluids

M. M. Stoican (Prisecaru)[1(✉)], N. Băran[2], D. Besnea[2], and A. Costache[2]

[1] Faculty of Mechanical Engineering and Mechatronics, University Politehnica of Bucharest, Bucharest, Romania
mirela.prisecaru@yahoo.com

[2] University Politehnica of Bucharest, Bucharest, Romania
n_baran_fimm@yahoo.com,
d_bes@yahoo.com, costache_adrian_1978@yahoo.com

Abstract. The present paper is a contribution to the theoretical and experimental researches carried out in the field of rotating machines with profiled rotors, which transports fluids.

Different forms of rotating pistons that transport the fluid from the suction to the discharge are presented.

The way in which the sealing between the two rotors is ensured, that in their rotational motion must be tangent all the time is analyzed. The fact that they are tangent leads to the delimitation of the low pressure area (suction) and the high pressure area (discharge); three constructive solutions of the rotating piston are presented and the most efficient solution is chosen.

Keywords: Rotating machine · Profiled rotor · Rotating piston

1 Introduction

The achievement of high performance rotating machines (pumps, fans, blowers) is topical.

The researches aims to build machines to ensure the transformation of the motor moment received from the shaft into useful effects, but with small energy losses.

These machines comprise rotating machines with profiled rotors [1, 2], at which the motor torque (M) is maximum during a complete rotation; this is because $M = F \cdot b \cdot \sin \alpha \;[N \cdot m]$, the angle α between the force (F) and the arm (b) is always equal to 90°.

The term rotating machine in the title of the paper refers to the fact that this machine can be used as:

- force machine if the suction pressure (p_1) is higher than the discharge pressure (p_2);
- working machine when $p_1 < p_2$.

G. I. Gheorghe (Ed.): ICOMECYME 2020, LNNS 143, pp. 1–11, 2020.
https://doi.org/10.1007/978-3-030-53973-3_1

Table 1 presents the classification of rotating machines with profiled rotors according to the purpose pursued and the adopted constructive solution [1].

Table 1. Classification of rotating machines with profiled rotors

Classification by purpose	Classification in terms of construction
Working machines	Pumps for driving fluids or with suspensions
	Fans for transporting gases or vapors
	Blowers for gas and vapor compression
Force machines	Hydraulic motor
	Pneumatic motor
	Steam engine or combustion gases

Rotating machines have the following advantages [3, 4]:

- The motor torque taken from the shaft is efficiently used to increase the fluid pressure from suction (p_1) to discharge (p_2);
- In the operation of the rotating machine there are no alternative rectilinear movements;
- The machine can carry clean fluids or with different particles: ash, sand, etc.;
- The machine has a high reliability in operation and various calculation programs have been made for its construction.

The advantage of this constructive solution is that the pumps can transports corrosive fluids, different acids, and other fluid substances that do not attack the material from which the rotors and casings are built.

The aim of this paper is to find the best shape of the profiled rotor. The rotor configuration can have different shapes; a contour of the rotor must be found so that the following conditions are met:

The losses by means of friction between the rotor and the casings must be as small as possible;

- Ensure a good seal between the rotors and between the rotors and the casings.

2 The Operating Principle and Constructive Solutions of the Rotating Machine with Profiled Rotors

The rotating machine will be analyzed as a rotating volumetric pump with profiled rotors; each rotor is provided with two rotating pistons made in the following variants:

Variant I: The rotating pistons have the form of rectangular blades;
Variant II: The rotating pistons are in the shape of an isosceles triangle;
Variant III: The rotating pistons have curvilinear shape.

The operating principle for the three constructive variants is similar.

a) Variant I: The Rotating Pistons Have the Form of Rectangular Blades.

- The movable part comprises the rotors (2) and (5), which are fixed on the shafts (7) and (9). On each rotor two rectangular blades are fixed that constitute the rotating pistons; the four blades separate the low pressure side (suction) from the high pressure side (discharge), (Fig. 1).
- The fixed part consists of two semi-cylindrical casings (1) and (4) provided with a suction connection (3) and a discharge connection (8). At a rotation of each rotor, two useful volumes are transported from suction to discharge.
- A useful volume is given by the product between the area of section (A) and the length (l) of the rotor (dimension perpendicular to the plane of the figure); the section area is between two successive blades and the lower casing (1).

$$V_u = 2 \cdot A \cdot l \ \left[m^3\right] \tag{1}$$

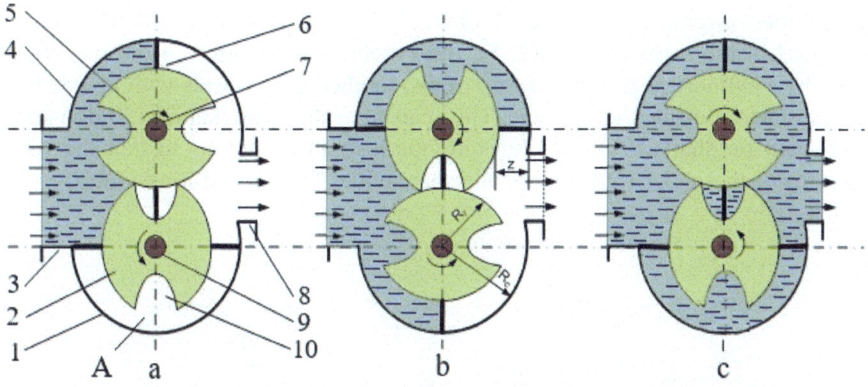

Fig. 1. Section through the rotating machine in variant I: 1 - lower casing; 2 - lower rotor; 3 - suction chamber; 4 - upper casing; 5 - upper rotor; 6 - rotating piston in the form of a rectangular blade; 7 - driven shaft; 8 - discharge chamber; 9 - driving shaft; 10 - cavity in which the piston of the upper rotor enters

The aspirated fluid (Fig. 1.a) is transported to the discharge and after 90° rotation of both rotors it reaches the situation in Fig. 1.b and later in Fig. 1.c. After a 180° rotation, the fluid in the useful volume V_u will be sent to the discharge chamber. At a complete rotation of the shaft (9) two such volumes will be transported from the suction to the discharge:

$$V_u = 2 \cdot \left(\frac{\pi R_c^2}{2} - \frac{\pi R_r^2}{2}\right) \cdot l \quad [m^3/rot] \tag{2}$$

where:

- R_c - Casing radius (Fig. 1.b);
- R_r - Rotor radius (Fig. 1.b).

The casing radius (R_c) is the sum between the rotor radius (R_r) and the piston height (z):

$$R_c = R_r + z \ [m] \tag{3}$$

Substituting the relation (3) into the relation (2) and taking into account that the machine has two rotors, the volumetric flow rate carried by the machine results:

$$\dot{V}_I = \pi l z(z + 2R_r) \cdot \frac{n_r}{30} \left[m^3/s\right] \tag{4}$$

For variants II and III the useful volume V_u will be smaller because the piston section is larger, as a result:

$$\dot{V}_I > \dot{V}_{II} > \dot{V}_{III} \left[m^3/s\right] \tag{5}$$

b) Variant II: The Rotating Pistons Are in the Shape of an Isosceles Triangle. The machine consists of two identical rotors (3, 4) of special shape, which rotate at the same speed inside some casings (2, 5); the synchronous rotation of the two rotors is ensured by a cylindrical gear consisting of two toothed gear wheels with inclined teeth, located inside or outside the machine. The gears have the same diameter of division and are mounted on shafts 7 and 8; they provide a rotation motion so that the rotating pistons (6) of the upper rotor enters into the cavities (10) of the lower rotor.

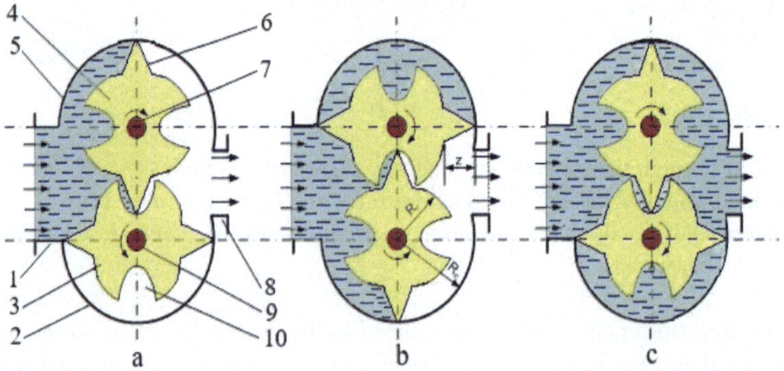

Fig. 2. The operating principle of the rotating volumetric machine in variant II: 1 - suction chamber; 2 - lower casing; 3 - lower rotor; 4 - upper rotor; 5 - upper casing; 6 - rotating piston; 7 - driven shaft; 8 - discharge chamber; 9 - driving shaft; 10 - cavity in which the upper rotor piston enters

In Fig. 3 the two gear wheels (7) are mounted outside the machine on the shafts (3) and (4); the lubrication of the cylindrical gear formed by these gear wheels is provided with the oil in the box (5).

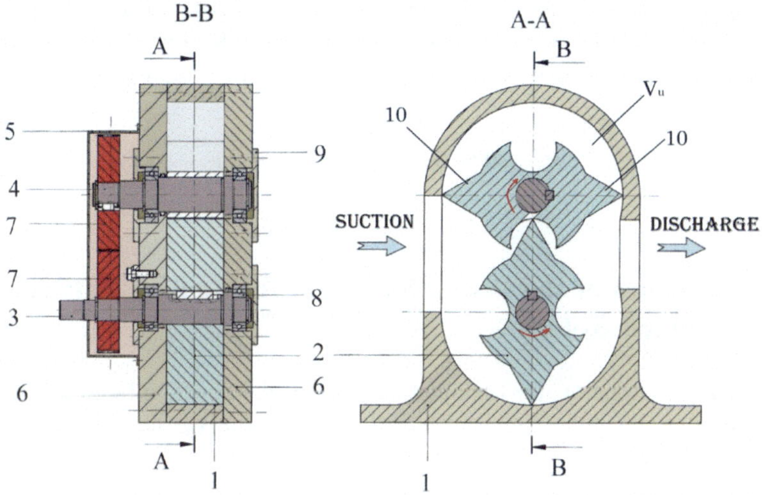

Fig. 3. Rotating volumetric pump with profiled rotors, 1 - oval casing; 2 - profiled rotor; 3 - driving shaft; 4 - driven shaft; 5 - oil box; 6 - side wall; 7 - gear wheels; 8 - bearing; 9 - bearing cover; 10 - rotating piston.

The fluid entering the suction chamber is transported to the discharge chamber regardless of its nature or composition [5, 6].

c) Variant III: The Rotating Pistons Have Curvilinear Shape.
Compared to variant II, here (Fig. 4) only the shape of the rotating pistons differs; the operation of the machine is similar to the one presented in variant I.

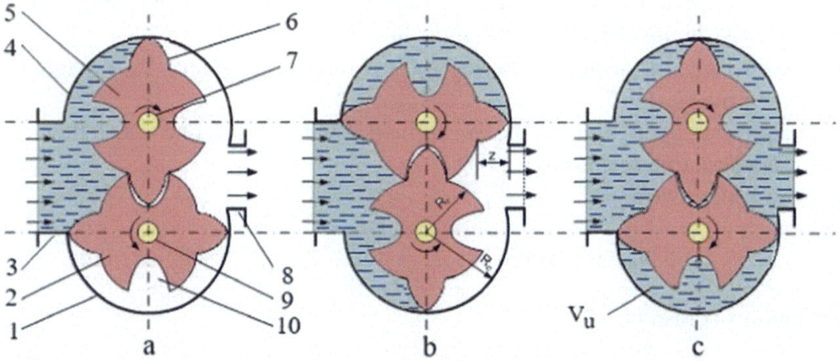

Fig. 4. Cross section through the rotating machine in variant III: 1 - lower casing; 2 - lower rotor; 3 - suction chamber; 4 - upper casing; 5 - upper rotor; 6 - rotating piston with curvilinear profile; 7 - driven shaft; 8 - discharge chamber; 9 - driving shaft; 10 - cavity in which the piston of the upper rotor enters

From Fig. 4 one can observe that the cross-sectional area of a rotating piston of curvilinear form is larger than of a triangular piston (see also Fig. 7); consequently the useful volume (V_u) will be smaller, so the volumetric flow rate will be lower:

$$\dot{V}_{III} < \dot{V}_{II} < \dot{V}_I \left[\frac{m^3}{s}\right] \quad (6)$$

3 The Influence of the Piston Shape on the Sealing Between Two Profiled Rotors

During a complete rotation, the delimitation in the radial direction between the suction area (low pressure) and the discharge area (high pressure) is ensured by the following contacts:

1 - between the sharp tip of the pistons and the casings (Fig. 2);
2 - between the sharp tip of the upper piston and the lower rotor (2).

The more perfect these contacts (the lower the interstices), the more the flow from the high pressure area to the low pressure area, the so-called "reverse flow" will have a lower flow rate, so the machine will have a higher volumetric efficiency.

Only point 2 will be analyzed below.

Details regarding the sealing area between the two rotors will be presented, for the three studied variants: I, II, III.

I. Variant I: The Rotating Pistons Have the Form of Rectangular Blades. From Fig. 5 one can observe that the upper rotor (4) through the piston in the form of a rectangular blade has a single contact point with the lower rotor (3).

This contact point denoted by K for a rotor of length l (dimension perpendicular to the plane of the figure) becomes a straight line.

In other words, the sealing between the two rotors is provided by a single contact area between the blade (7) and the lower rotor (3).

To achieve a good seal in the mentioned contact area, one must have a high precision construction technology [7–9].

For the given sizes:

$$l = 0.05\,[m];\; z = 0.03\,[m];\; R_r = 0.05\,[m];\; n_r = 300\,[rot/min] \quad (7)$$

from relation (4) results:

$$\dot{V} = \pi \cdot 0.05 \cdot 0.03(0.03 + 2 \cdot 0.05) \cdot \frac{300}{30} = 0.006123\,[m^3/s] \quad (8)$$

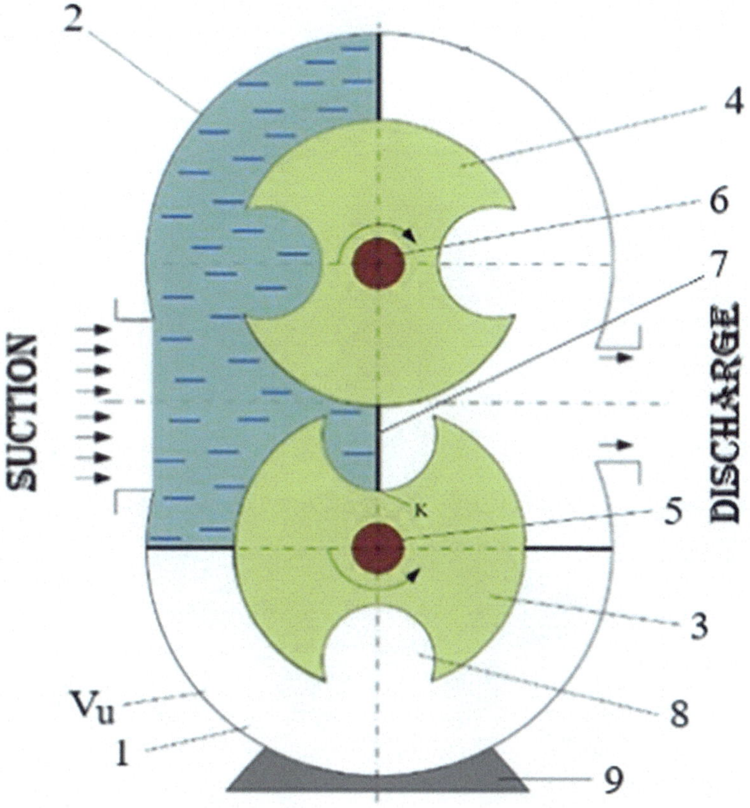

Fig. 5. Cross section through the rotating working machine: 1 - lower casing; 2 - upper casing; 3 - lower rotor; 4 - rotor upper; 5 - driving shaft; 6 - driven shaft; 7 - rotating piston of blade shape; 8 - cavity; 9 - machine support

II. Variant II: The Rotating Pistons Are in the Shape of an Isosceles Triangle. Figure 6 shows that between the upper rotor (2) and the lower rotor (1) there is only one contact point noted with M; if the piston (3) is built with a larger base, it will lock in the cavity in the rotor (1) [10–12].

Also as in variant I along the rotor length (l) the contact point (M) will describe a contact line between the rotating piston (3) and the lower rotor (1).

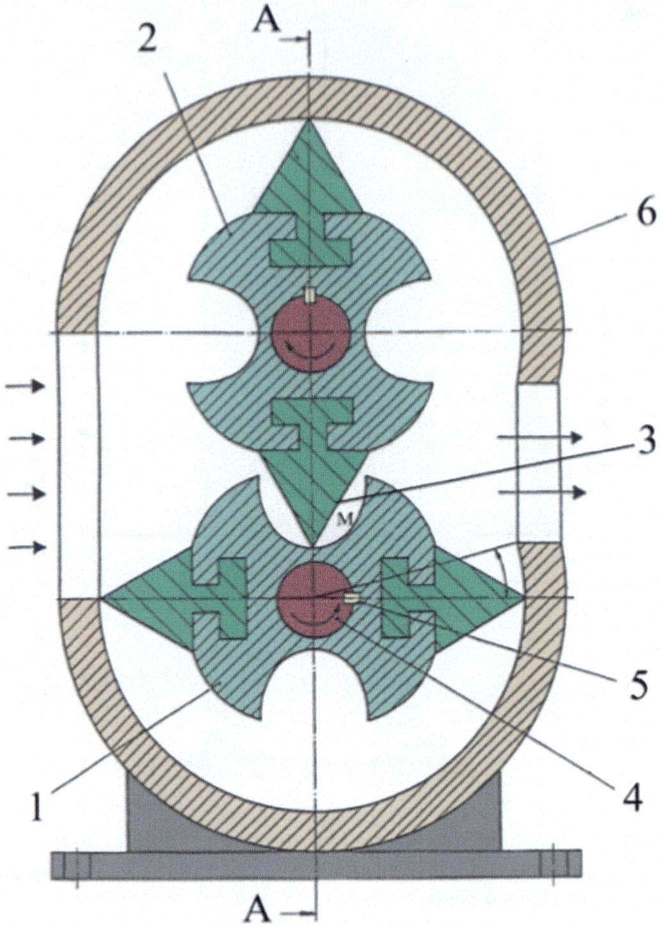

Fig. 6. Cross section through rotors with triangular shaped pistons: 1 - lower rotor; 2 - upper rotor; 3 - triangular piston; 4 - shaft; 5 - rectangular wedge; 6 - machine casing

III. Variant III: The Rotating Pistons Have Curvilinear Shape.

To ensure better sealing between the two rotors, the piston profile will not have the shape of a triangle. The profile will be made of two curves symmetrical to the *oy* axis; these curves are not part of a circle, but are mathematically established, the shape of the contour of the curvilinear piston being given by points.

The curved shape of the piston ensures better contact between the rotating piston and the cavity of the adjacent rotor.

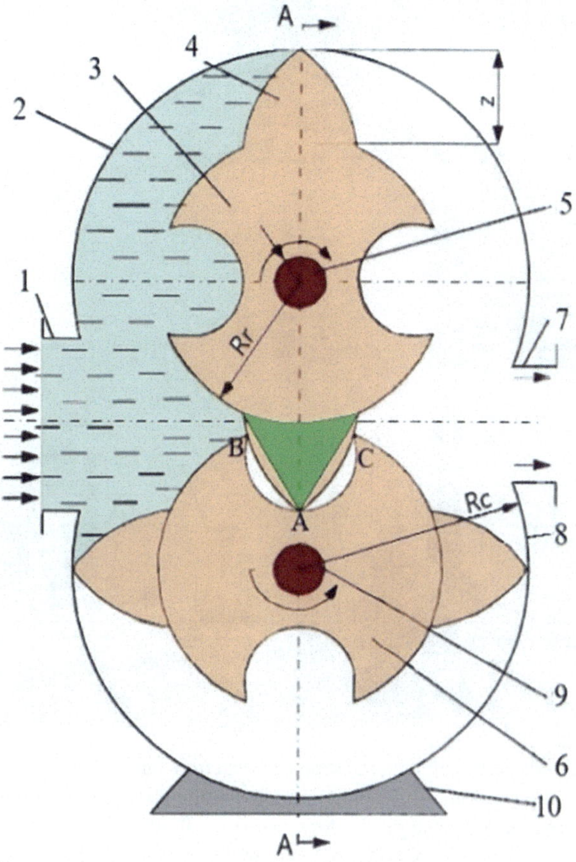

Fig. 7. Cross section through the rotating machine: 1 - fluid suction chamber; 2 - upper casing; 3 - upper rotor; 4 - rotating piston; 5 - driven shaft; 6 - lower rotor; 7 - fluid discharge chamber; 8 - lower casing; 9 - driving shaft; 10 - machine support.

Figure 7 shows:

- The triangular profile of the rotating piston;
- Curvilinear profile of the rotating piston.

If the piston is in triangular shape, there is only one contact line (area A) between the piston and the cavity in the adjacent rotor.

If the piston is curvilinear, in Fig. 7 one can observe that there is three contact lines: B, A and C; as a result, fluid losses between suction and discharge through “reverse flow” will be reduced.

Both profiles were mathematically established, i.e. the coordinates (x_i, y_i) [13] and the manufacturing technology were specified.

Figure 8 shows a concrete solution of the rotating machine made in the laboratory and its framing in the experimental installation.

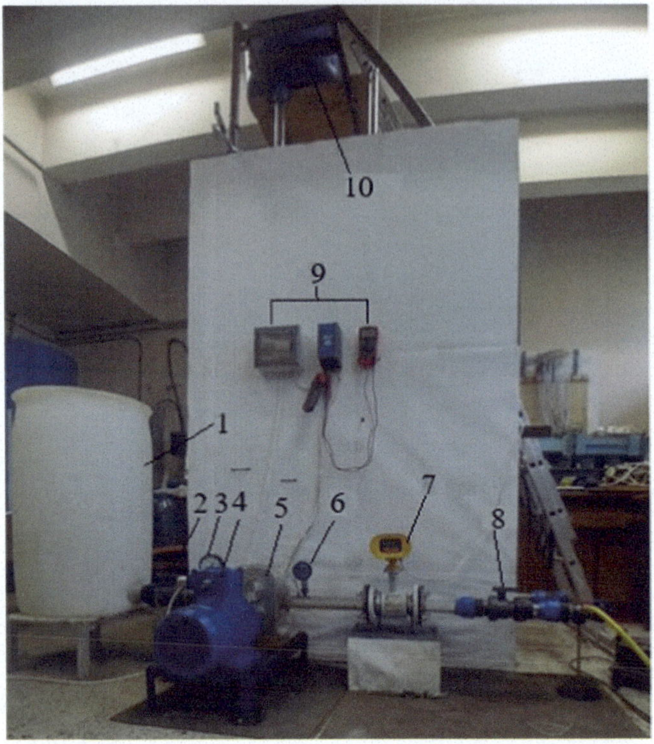

Fig. 8. View of the experimental installation: 1 - suction tank; 2 - tap Dn 60 Pn 2 bar; 3 - pressure gauge at pump suction; 4 - the electric motor of the pump; 5 - rotating volumetric pump; 6 - pressure gauge at pump discharge; 7 - flowmeter; 8 - fluid flow rate control valve; 9 - panel with measuring devices (frequency converter, voltmeter, ammeter); 10 - discharge tank.

4 Conclusion

a) For a certain required volumetric flow rate, a certain length of the rotor is chosen depending on the execution technology; then the rotor radius is chosen and the rotating piston height results.
b) The change of the volumetric flow rate of the rotating machine is done by changing the speed of the engine driving the machine.
c) This machine can carry multiphase fluids, viscous fluids; as a result, it can be used in the fields of energy, petrochemical, food, agriculture.
d) The constructive solution for the rotating machine was designed and built in the laboratory of the Faculty of Mechanical and Mechatronics Engineering, the Department of Thermotechnics, Engines, Thermal and Refrigerating Equipment's of University Politehnica of Bucharest.
e) The experimental tests were analyzed and following the results of the experimental measurements it was found:

- The analysis of the three variants shows that variant III is the best.
- This presented constructive solution ensures a better sealing between the high pressure part (discharge) and the low pressure part (suction) of the working machine (pump, blower).

In conclusion, between the three analyzed constructive variants, from the sealing point of view and reducing the flow of the "reverse flow", the variant of the profiled rotor with pistons of curvilinear form is chosen (variant III).

References

1. Băran, P.N., Răducanu, A.O.: Bases of Technical Thermodynamics, Technical Thermodynamics (in Romanian). Politehnica Press Publishing House, Bucharest (2010)
2. Motorga, A.: Influence of constructive and functional parameters on the performances of rotating machines with profiled rotors, Ph.D. thesis, Faculty of Mechanical Engineering and Mechatronics, Politehnica University of Bucharest (2011)
3. Detzortzis, A.: Influenţa arhitecturii rotoarelor asupra performanţelor compresoarelor volumice rotative cu rotoare profilate, Teză de doctorat, Universitatea Politehnica Bucureşti (2014)
4. Băran, N.: Maşini termice rotative de lucru. Matrix Rom, Bucureşti (2003)
5. Hawas, M.: The influence of fluid viscosity on the performance of rotating machine with profiled rotors, Ph.D. thesis, Universitatea Politehnica din Bucureşti, Romania (2015)
6. Băran, N., Căluşaru, I., Detzortzis, A.: Research regarding the testing of a new type of rotating machine with profiled rotors. J. Mater. Sci. Eng. A **2**(3), 372–376 (2012)
7. Cristea, V., Creţa, G., Ivan, D., Ardeleanu, P.: Etanşări. Tehnică, Bucureşti (1973)
8. Ţurcanu, C., Ganea, N.: Pompe volumice pentru lichide. Tehnică, Bucureşti (1980)
9. Besnea, D., Donţu, G.O., Alexandrescu, N.: Tehnologii de fabricaţie asistate de calculator pentru execuţia unor componente mecatronice. Printech, Bucureşti (2008)
10. Băran, N., Besnea, D., Detzortzis, A., Bărascu, A.: Manufacturing technology of a new type of profiled rotor used by a rotating volumetric pump. Proc. Manuf. Syst. **7**(2), 105–110 (2012). ISSN 2067-9238
11. Băran, N., Duminică, D., Besnea, D., Detzortzis, A.: Theoretical and experimental researches regarding the performances of a new type of rotating machine with profiled rotors. Adv. Mater. Res. **488–489**, 1757–1761 (2012)
12. Băran, N., Căluşaru, I., Detzortzis, A.: Research regarding the testing of a new type of rotating machine with profiled rotors. J. Mater. Sci. Eng. **2**(3), 372–376 (2012). ISSN 2161-6213
13. Băran, N., Besnea, D., Motorga, A.: Elements of computing the architecture and manufacturing technology for a new type of profiled rotor. In: Proceedings International Conference, 6th Workshop on European Scientific and Industrial Collaboration on promoting Advanced Technologies in Manufacturing, WESIC 2008, Bucharest, pp. 233–241 (2008)

Mechanical Behaviour Analysis of Snap Joints for Haptic Evaluation

Ciprian Ion Rizescu(✉) and Dana Rizescu

University POLITEHNICA of Bucharest,
Splaiul Independentei 313, Bucharest, Romania
ciprian.rizescu@upb.ro

Abstract. A very simple and cheap joining technique is using the snap joints. They are a characteristic type of joint used for polymer materials, since the material properties (flexibility) and moulding potential of polymers are particularly conducive to this kind of connection. The latest manufacturing processes take into account the additive technologies using 3D printers and different materials. The authors developed an experimental setup for testing different materials, such as ABS and PLA, different fill factors, friction coefficients between parts and different geometries for snap joints. Also a FEM simulation of mechanical behaviour, developed in solidworks presents the differences between solutions and underline the advantages and disadvantages of each snap joint solution. The experimental setup allows fast changing of snap joints parts. This research aims the study of behaviour of 3D Printed machine elements like: springs, bearings, clutches, gears, bellows, diaphragms, bushes, brakes, sliders, etc. developed previously by authors.

Keywords: Snap joints · Fastening connections · FEM simulation

1 Introduction

1.1 Snap Joints

Locking devices are increasingly being used as joints for pairing different parts. The application of the fastening connections has become widespread in the automatic assembly.

Two aspects are important regarding the design of the fastening joints, the joining forces during assembly and the forces during disassembly. The first mentioned forces must be applied for insertion, and the latter are responsible for the fastening function.

In this paper we present the modelling and simulation of insertion and extraction of fast fixing elements. Therefore, we are able to predict the forces and moments acting in the assembly time and the maximum force that can be carried by the fast assembly locking arm.

Also, in the modern industry, the methods of rapid assembly, of the different components and constructive parts, based on locking mechanisms, are increasingly used. This fact is in accordance with the new design concepts, in which the saving of raw materials and materials plays an important role, since it is realized the replacement

G. I. Gheorghe (Ed.): ICOMECYME 2020, LNNS 143, pp. 12–19, 2020.
https://doi.org/10.1007/978-3-030-53973-3_2

of some elements of metallic assembly (screws, rivets) with pieces of plastic materials. Areas of use of plastic mass locking mechanisms are very varied. These include: assembly of housings, connectors, in within the C&C office systems, audio-video equipment, on-board car equipment, etc. The advantages offered by these elements of assembly, in fact locking mechanisms or elastic locking, are: simple construction, unpretentious execution technologies, easy handling, weight and size low and a lower price than the metal assembly elements that impose special conditions of execution and assembly. Two examples of products equipped with such locking mechanisms (a fuse safety socket and a circular connector body) are illustrated in Fig. 1.

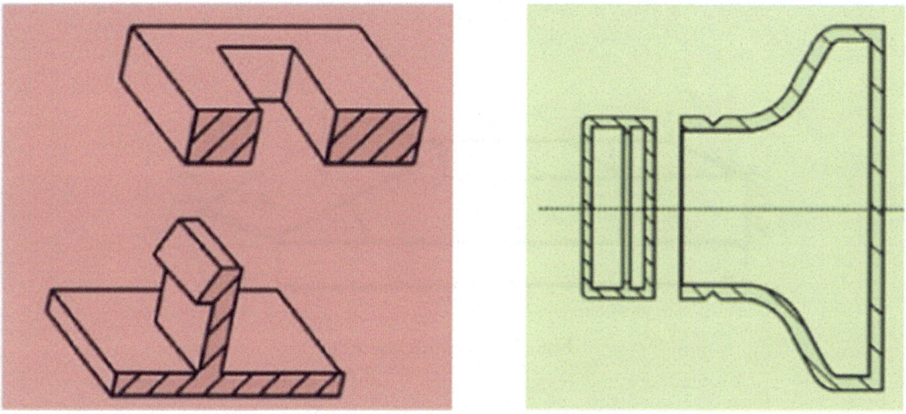

Fig. 1. A fuse safety socket (left) and a circular connector body (right).

1.2 Common Snap-Fit Connections

There are many grouping criteria for these elements. Generally, there are two main types of snap-fit connections that are appropriate for 3D printing: cantilever and annular. The cantilever is the most common snap-fit joint and consists of a protrusion (some type of bead or hook) at one end of the beam and a structural support at the other end. This protrusion is inserted into a cut-out or slot and deflects upon insertion. Once fully inserted, the protrusion bends back locking the connection into place.

Cantilever snap-fits are easy to design and intuitive for the user during assembly and disassembly. In many cases, it's the cheapest way to join 2 parts together. The annular snap-fit utilizes hoop-strain to hold a pressed part in place. Common examples of annular snap-fits are bottle and pen caps.

To avoid premature destruction of these locking mechanisms, due to couplings and repeated decoupling, but also for a correct operation of the device in which they are mounted, is it is necessary to know the parameters that influence their functioning state. Their parameters are: mounting force, coupling force (locking), which requires the elastic element to bend, coefficient of friction, the parameters related to the geometry of the coupling parts: length, width, angle of inclination, thickness, radius of curvature. Knowing these parameters is possible through research experimental, by reproducing the operating conditions on a test and measurement stand.

2 Mechanical Behaviour Study

The experimental study for determining the behaviour of the coupling parts aims to compare the theoretical results with the experimental ones and, eventually, to improve the theoretical relationships, which allow to determine the geometry of the elastic element, as well as the experimental verification of the connection mounting force (F) - coupling force (locking) (Q).

Most of the coupling parts have the shape of a simple hook or ratchet, the protrusion that actually achieves the joint can have smaller or larger inclination angles. Among the most used forms of the coupling parts, the one presented in Fig. 2 was retained.

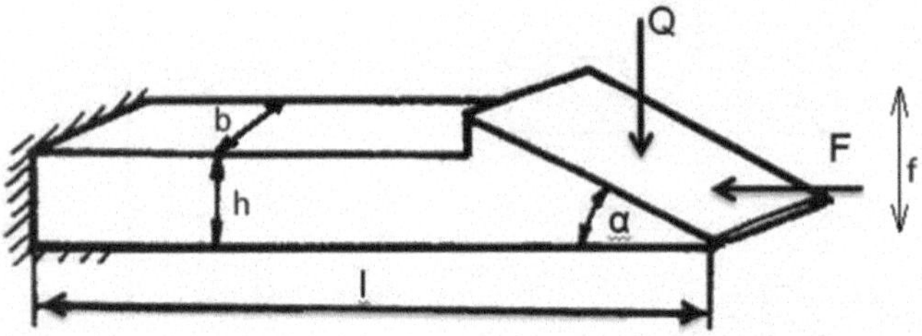

Fig. 2. The locking arm

It should be noted that the elastic elements of the coupling parts are required for bending, mostly, but there are also situations when the coupling is done by twisting the elastic element.

For straight and circular coupling parts, the joint can be made by inserting the part by pressing into the comprehensive part, or vice versa, by moving the base part (comprehensive) towards the part to be covered. The deformation f, represents the displacement of the free end of the elastic element in the joining process and is dependent on the geometry of the coupling parts (h, b, l, α), on the type of plastics used, implicitly on the specific elasticity and elongation of material.

The calculation relation between the coupling force (locking) Q and the mounting force F is given by the relation (1) [1]:

$$F = Q \cdot \frac{\mu + tg\alpha}{1 - tg\alpha} \tag{1}$$

where μ is the friction coefficient between parts, α-locking angle. The force Q can be calculated with the relation (2) [1]:

$$Q = \frac{b \cdot h^2}{6} \cdot \frac{\varepsilon \cdot E}{l} \tag{2}$$

where: l - arm length; b - arm width h - arm thickness; ε is the specific elongation, and E is the longitudinal modulus of elasticity (Young modulus).

There were consider two materials: PLA (Polylactic acid) and ABS (Acrylonitrile butadiene styrene) for both snap joint parts [2]. The authors developed also an experimental setup for testing different materials, such as ABS and PLA, different fill factors, friction coefficients between parts and different geometries for snap joints [3–5]. The authors developed a FEM simulation of snap joints in Solidworks environment in order to study the assembling behavior of locking arm for different locking angles, materials, fill factors. In Table 1 there are presented the FEM simulation results for these two materials and three locking angles, α.

Table 1. Comparison for different locking angles and two materials.

α [°] / Material	Arm displacement	Stress Analysis	Computed coupling force
α=15° / PLA	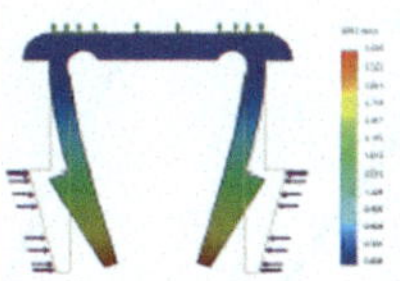	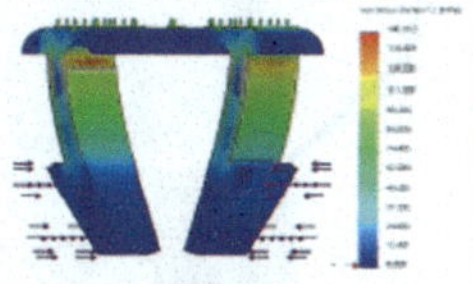	Q=37.5N
α=15° / ABS	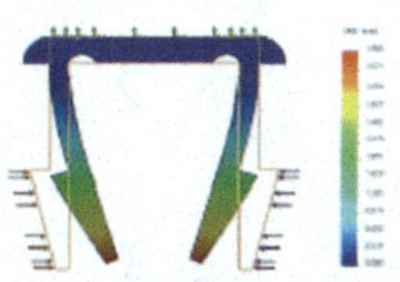	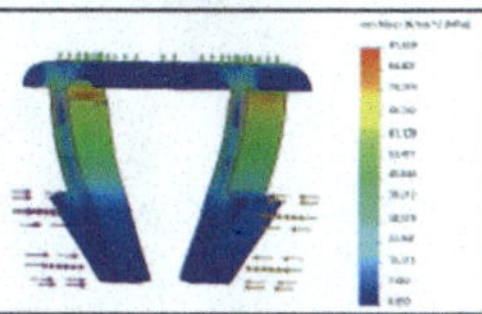	Q=62,25N
α=30° / PLA	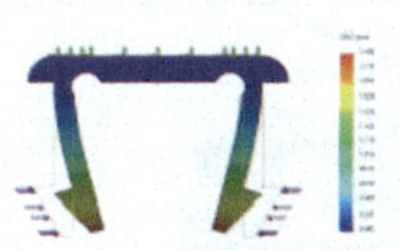		Q=47N
α=30° / ABS	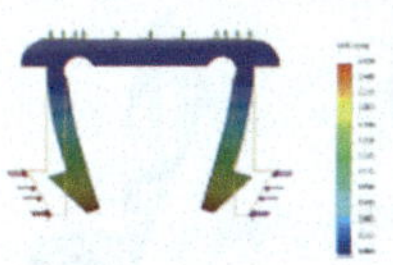		Q=82N
α=45° / PLA			Q=56,25N
α=45° / ABS			Q=98.43N

3 Experimental Setup

The experimental setup (see Fig. 3) consists of subsequent elements: 1 - deflection measuring system, 2 - Imada force transducer, 3 - plate for the snap joints, assembling mode, 4 - the x, y positioning table. Maximum testing force is 500 N with 0.01 N accuracy [6]. The Imada transducer is connected to PC for record the insertion force and time. The testing end is of a cylindrical shape, with a plane surface pushing the joint part. The pressing process of the cantilever joint part is presented above (see Fig. 3). The joints parts: cantilever joint parts and plate parts are manufactured from ABS – yellow and PLA – violet (see Fig. 4)

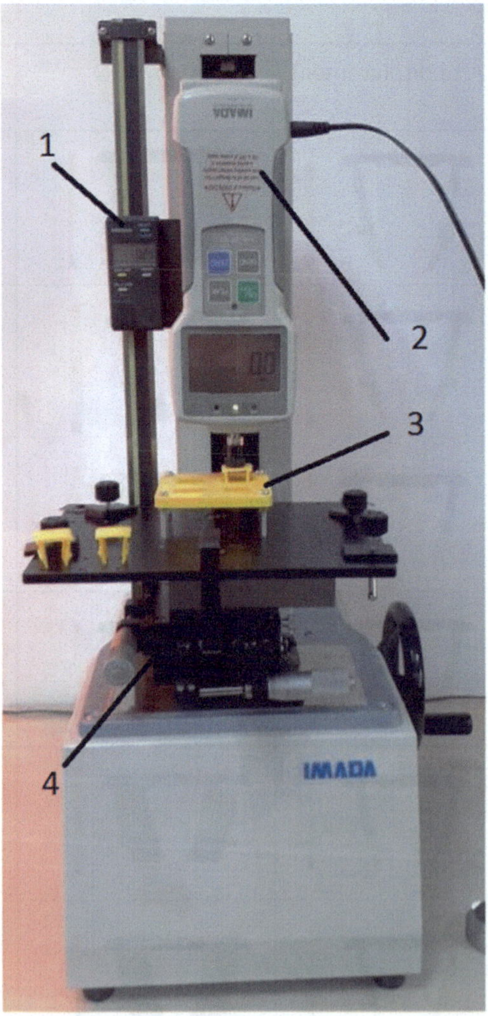

Fig. 3. Vertical test stand from Hans Schmidt with Imada Transducer 500 N.

Fig. 4. Plate parts (left) and cantilever parts (right).

Each cantilever part was paired with its corresponding rectangular hole plate with respect to the locking angles, α. These angle were considered of 15°, 30°, 45° for both: the cantilever parts and for plates. There is a plane rectangular hole in the plates with locking angle $\alpha = 90°$. For each cantilever, with its locking angle, α, the insertion process was developed as it is shown in Table 2.

Table 2. Comparison for different locking angles and two materials.

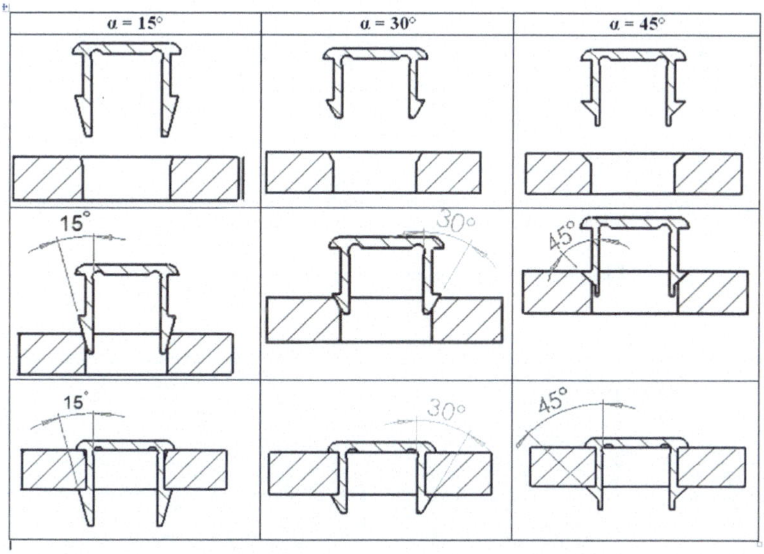

The insertion forces and corresponding displacements of Imada transducer were recorded during cantilever inserting operation. This operations were developed for both materials: ABS and PLA [7, 8]. The results are presented in Figs. 5, 6 and 7 with respect to the plate material. The fill factor for all parts of snap joints is 30%. In Fig. 5 the plate is manufactured from PLA and cantilever from ABS and PLA. When the cantilever is manufactured from PLA the insertion forces exceed 60N (see Fig. 5 and Fig. 6). When the cantilever is manufactured from ABS the insertion forces are lower than 20 N.

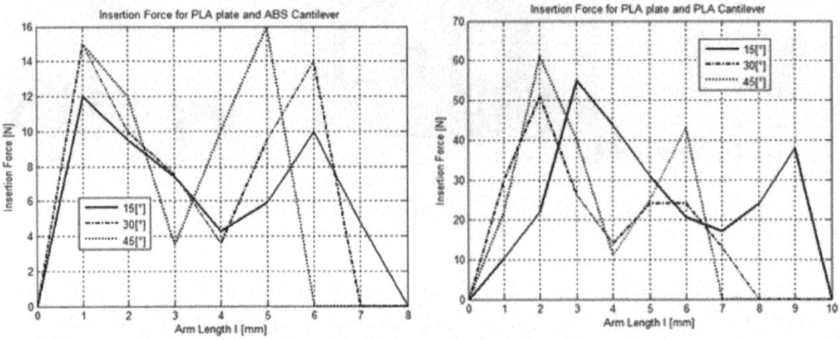

Fig. 5. The PLA plate and ABS, PLA cantilever

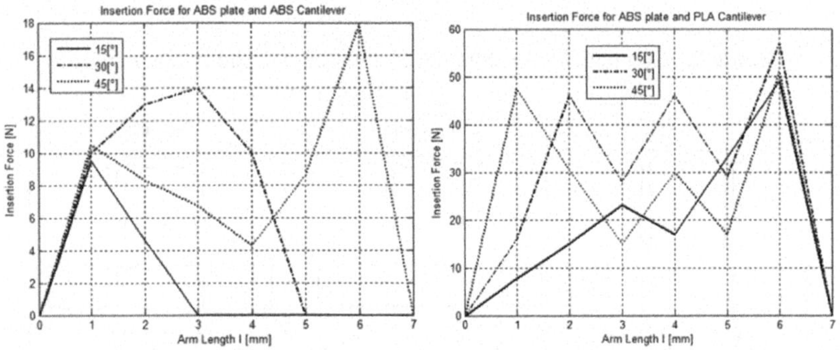

Fig. 6. The ABS plate and ABS, PLA cantilever

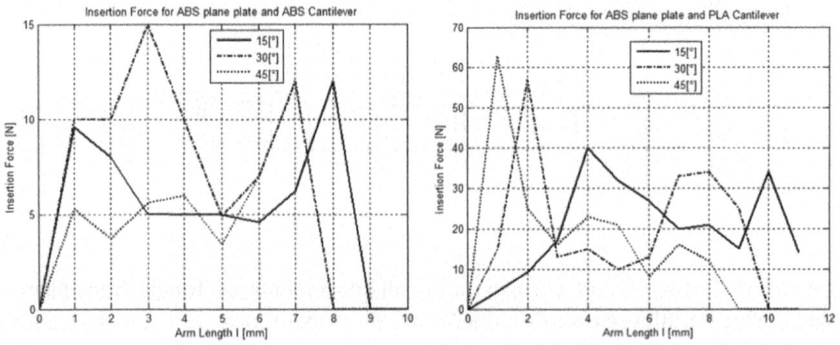

Fig. 7. The ABS plane plate and ABS, PLA cantilever

When the locking angle, $\alpha = 90°$, the plate is called plane plate and the insertion forces have the same behaviour as it was described above (see Fig. 7). For ABS plane plate and PLA cantilever the forces exceed 60 N. The forces for ABS plane plate and ABS cantilever are lower than 20 N.

Using Eq. 1 for PLA cantilever and: $\mu = 0.32$; $E = 3500\ N/mm^2$, $\varepsilon = 5\%$, $l = 15$ mm, $b = 15$ mm, $h = 1.5$ mm, $\alpha = 15°$ the insertion force is 42,75 N.

4 Conclusions

The authors developed an experimental setup for testing different materials, such as ABS and PLA, different fill factors, friction coefficients between parts and different geometries for snap joints.

There was developed a FEM simulation of mechanical behaviour of the snap joints in solidworks. Even there were some difficulties to find out some mechanical characteristics like: elasticity modulus, density for fill factor 30%, friction coefficient the simulated results are near the experimental ones: the insertion force was about 37.50 N in simulation and 40 N in experimental tests, for the locking angle $\alpha = 15°$ and PLA cantilever. Using Eq. 1, the theoretical insertion force is 42.20 N. These differences show that the experimental tests are in good agreement with simulations and theoretical studies.

References

1. Rizescu, C.I., Udrea, C., Panaitopol, H.: Laboratory Textbook - Fundamentals of Mechatronics Instruments (in Romanian), pp. 69–74. Printech, Bucharest (2000)
2. Berce, P., Balc, N., Caizar, C., Pacurar, R., Sever Radu, A., Bratean, S., Fodorean, I.: Manufacturing Technologies by Adding Material and Their Applications (in Romanian). Romanian Academy Publisher, Bucharest (2014)
3. Moraru, E., Dontu, O., Besnea, D., Constantin, V.: Study and realization of prosthetic dental models by additive technologies. IOP Conf. Ser. Mater. Sci. Eng. **444**, 1–8 (2018)
4. Besnea, D., Gheorghe, I.G., Dontu, O., Moraru, E., Constantin, V., Moga, I.C.: Experimental researches regarding realization of wastewater treatment elements by means of modern technologies. Int. J. Mechatron. Appl. Mech. **4**, 61–65 (2018)
5. Moraru, E., et al.: Fabrication technologies of aeration systems for the ecological treatment of wastewater. In: Gheorghe, G.I. (ed.) ICOMECYME 2019. LNNS, vol. 85, pp. 133–141. Springer, Cham (2020). https://doi.org/10.1007/978-3-030-26991-3_13
6. https://www.hans-schmidt.com/en/produkte/vertical-manual-test-stands/
7. http://www.ampolymer.com/SDS/PolylacticAcidSDS.html
8. https://www.makeitfrom.com/material-properties/Acrylonitrile-Butadiene-Styrene-ABS

Rapid Prototyping Boost in Research and Development

Karolina Macúchová[✉], Milan Melichar, Pavel Crha, and Jan Heřmánek

HiLASE Centre, Institute of Physics CAS, Za Radnicí 828, 25241 Dolní Břežany, Czech Republic
karolina.macuchova@hilase.cz

Abstract. In this article we demonstrate benefits that rapid prototyping methods bring to a modern hi-tech laboratory. Research and development facilities, such as the HiLASE center, handles a huge amount of custom components that needs to be fitted and adjusted on site.

The HiLASE center focuses on innovative laser technologies. Due to fast speed of research we have to speed up the design process as well. Production method based on in-house fast and cheap 3D printing enables effective testing of custom design solutions and avoids time delays when compared to outsourced common subtractive manufacturing technology. For numerous special design applications the additive 3D printing methods offers manufacturing capabilities that cannot be reached otherwise.

Keywords: Laser facility · Technology · Manufacturing · Rapid prototyping · FFF · FDM · 3D printing · Opto-mechanical design · HiLASE

1 Introduction

HiLASE is a user scientific facility developing more than twenty laser systems for scientific and industrial applications, with output parameters ranging from a few picosecond pulses with energies of 5 mJ–0.5 J and repetition rates of 1–100 kHz (thin disk technology) to systems with 100 J output energy in nanosecond pulses with a repetition rate of 10 Hz (multi-slab technology) [1]. One of the key features of the HiLASE facility is its ability to deliver variety of these laser beams to experimental stations located in different user laboratory.

In the last few years, the use of additive technologies has found its way to all branches of industry. So it is no surprise that even the research institutions, like the HiLASE Center, has decided to include this technology in its production processes. From the beginning, we did not expect such a good yield from implementing 3D printing technology at the HiLASE center. But over time it proved to be the right move. The main advantage of the 3D printing technology is fast and cheap manufacturing of mostly good enough lightweight parts.

G. I. Gheorghe (Ed.): ICOMECYME 2020, LNNS 143, pp. 20–25, 2020.
https://doi.org/10.1007/978-3-030-53973-3_3

2 Design Process

At HiLASE, we use the FFF additive technology (Fused Filament Fabrication). This technology perfectly meets the needs of the HiLASE design team with its various versatility of production. Combining the development of the most powerful lasers in the world with the most basic 3D printing might seem strange at the first sight. But in the scope of time we have far succeeded to gain the most of benefits that FFF printing method offers.

2.1 Boosting the Design Process

The nature of a frontier technology research institute poses a lot of design challenges for the design team engineers. It happens frequently during the first stages of a design life-cycle that the initial design assignment and requirements changes several times so much, that it is easy to start from the scratch all over again.

In large scale, we combine the purchased components with the in-house made. Mostly due to lack of compatibility or lack of design features of the off-the shelf components we are driven to custom solutions. The design team has to usually implement and connect two and even more different systems to work together. This is the moment, when the 3D printing technology proves to be a very fast and effective tool that ease the design work.

2.2 Eliminating the Idle Time

By using 3D printing, we managed to significantly speed up the development process. Design parts, which were previously produced by the common subtractive methods, were delivered 4–6 weeks after the final approval of the design. Such long waiting times are very unfavorable for the development of new devices. All modifications take a lot of time. With the 3D print we are able to get hold of the parts within hours after design approval. So the parts can be assembled the second day after the completion of the 3D design. With such a high speed of prototype production, we are able to solve requests for design changes without major delays.

This gained flexibility proved very useful during user experiments at the HiLASE facility. Each user experiment is unique and therefore carefully planned several months in advance. If an initial analysis shows there is a compatibility error, e.g. wrong size of a sample holder, it is not possible to wait 4 weeks for a new one. In such cases, the 3D printing offers the instant solution. Within few hours, we are able to produce the necessary part and go on with preparation of the experiment.

The financial benefits of 3D printing cannot be overlooked either. The printed part is many times cheaper than a part produced by conventional methods. If we add the price for a delayed or cancelled experiment, the price difference would be higher by several orders. In the sum, whether we produce a part for € 100 or € 10 does not really matter, when the equipment time of device costing a million Euro is put on idle.

3 Design Cases

We will describe several custom design cases, which were solved with the 3D printing method.

3.1 Connecting the Imperial with the Metric System

The equipment used in the HiLASE Center comes from all over the world. It is no exception that parts made in the USA, Germany and England are mounted together within a single laser device. The already mentioned design flexibility enables connecting different parts originally designed in Imperial and Metric systems together. Such a converting part fits the right dimensions of both systems (see Fig. 1 and Fig. 2). The printed parts can also be used to unify the threads. Sufficient accuracy and rigidity can be achieved by smart selection of the right material and proper print settings. By setting the shape and density of part infill we get components with tunable features.

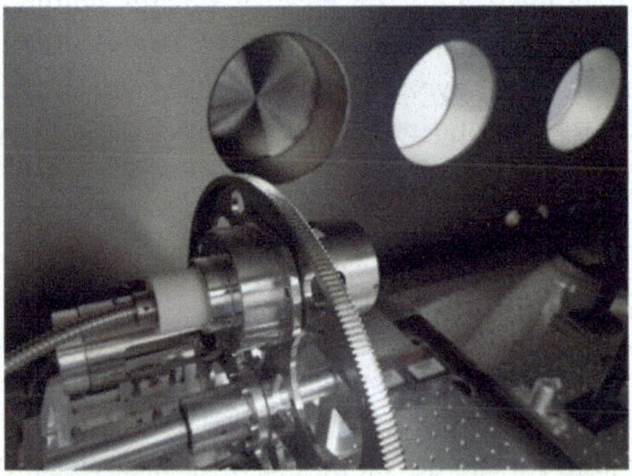

Fig. 1. Threads unifier mounted in an experimental chamber

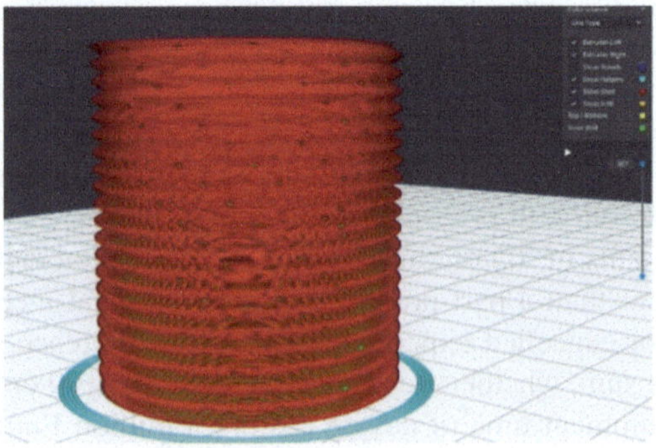

Fig. 2. Sliced connector ready to be printed

3.2 Rapid Prototyping

The impact of manufacturing a complex shaped parts is marginal for an additive technology compared to a conventional production. Additive technologies allow exploring new ways of design.

For example we can easily combine materials within one part. See Fig. 3 for a part designed for optics polishing. This polishing head is made directly from two materials which results in better its better working performance. It is obvious that new parts, not limited by normal manufacturing processes, can be better adapted to their purpose.

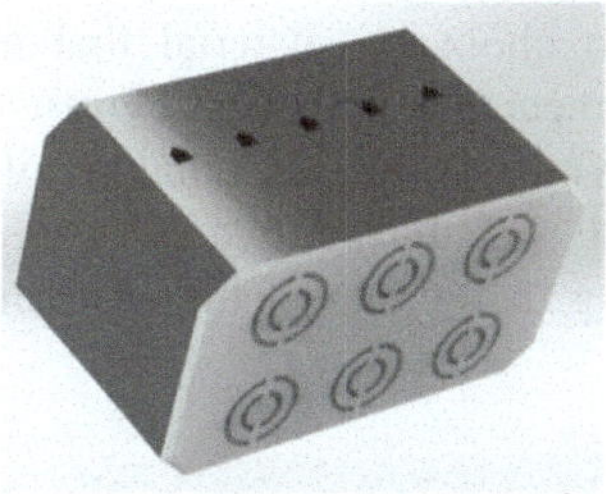

Fig. 3. Two material composite tool for optics polishing

3.3 Large Parts

Optical components of the laser systems are sensitive to vibration. The opto-mechanical parts, stands and holders therefore have to be stiff. The performance stability of these laser sub-systems cannot be compromised by using plastic parts [2]. But still the 3D printing technology can be used as a first step in the design process to check the design viability. The complexity of our laboratory equipment and space limitation pose a not negligible risk of collision with other components. To avoid this risk, a cheap plastic dummy part is produced before ordering the production of an expensive part. This cheap dummy part is mounted to the system and thus we verify that the design is correct. If a collision occurs, the part is redesigned at less cost and time consumption.

Fig. 4. Stage holder

On Fig. 4 there is an example of a prototyped stage holder. It was designed on site in the laboratory according to imminent spatial restrains. It was printed and mounted within few days to keep a deadline for a user experiment. When the design concept proved its worth, the stage holder was manufactured from Aluminum and replaced the plastic one. The vibration stability of the plastic holder was lower than that made of metal but still good enough to meet the requirements.

3.4 Vibration Attenuators

We have a variety of practices to follow when designing a special use component. Apart from the bare design features, there is a large selection of materials to use with 3D printing method. We can choose a material that meets the mechanical and/or chemical requirements. The mechanical properties can be tuned by designing infill spatial structure of the part. We can print parts that significantly reduce the amount of transmitted vibration as well. We have already designed in-house rubber shock-absorbers and sealing. See Fig. 5 for one of these attenuators that is placed under a laser device.

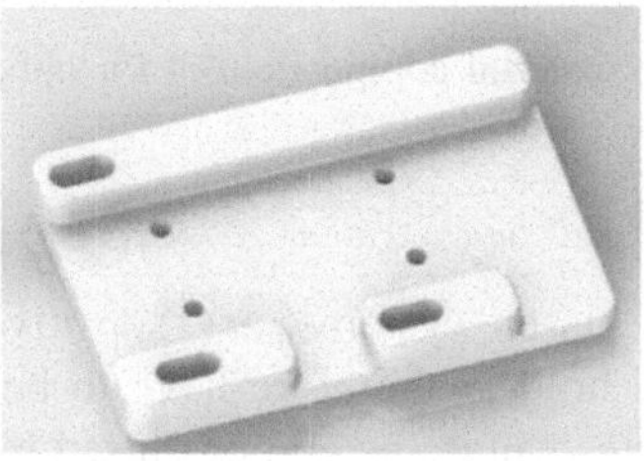

Fig. 5. Rubber shock absorber

3.5 Fiducials

For optics alignment purposes we design our custom fiducials, which are applicable for different types and shapes of opto-mechanical holders. See Fig. 6 for a selection of samples. These can be produced in redundancy and help us save time during the assembly of laser systems.

Fig. 6. Adjustment tools

4 Conclusion

FFF is considered as future of manufacturing technology owing to its versatility and ability to convert an idea into real products at a faster pace as compared to other conventional processes [3]. Despite so many promising features, 3D printing method has limitations mostly originating from material properties. It still lacks in quality of surface finish and strength. Plastic material is generally not suitable for parts that require high rigidity and vibration stability. But by means of good design supported by smart print infill settings we can obtain very nice results.

We have started with one FFF printer and currently we have four. We bought two printers with multiple extruders, enabling the production of one part from multiple materials. By combining two materials, the part can acquire unique properties or different colors. The demand for printed parts is still growing and in the future it will be necessary to purchase additional equipment. The number of printed parts in 2019 exceeded 700 pieces. If everything continues at the same pace, we expect to reach the number of 1000 parts in 2020.

Acknowledgement. This work was financed by the Ministry of Education, Youth and Sports of the Czech Republic (Programme NPU I Project No. LO1602).

References

1. Divoky, M., Smrz, M., Chyla, M., Sikocinski, P., Severova, P., Novak, O., et al.: Overview of the HiLASE project: high average power pulsed DPSSL systems for research and industry. High Power Laser Sci. Eng. **2** (2014). https://doi.org/10.1017/hpl.2014.16
2. Aliheidari, N., Tripuraneni, R., Hohimer, C., Christ, J., Ameli, A., Nadimpalli, S.: The impact of nozzle and bed temperatures on the fracture resistance of FDM printed materials. Proc. SPIE, Behav. Mech. Multifunct. Mater. Compos. **10165**, 1016512 (2017). https://doi.org/10.1117/12.2260105
3. Shubham, P., Sumit, J.: Optimization of process parameter to improve dynamic mechanical properties of 3D printed ABS Polymer using Taguchi method. In: ARSSS International Conference, Goa (2018)

Behavior of Composite Material Instrumented by Optical Fiber

R. El Abdi[1(✉)], V. Chean[1], H. Ramezani[2], P. Casari[2], and F. Jacquemin[2]

[1] Univ. Rennes - CNRS, Institut de Physique de Rennes, UMR 6251, 35000 Rennes, France
relabdi@univ-rennes1.fr

[2] LUNAM Université - Université de Nantes - Centrale Nantes, Institut de Recherche en Génie Civil et Mécanique, UMR CNRS 6183, BP 420, 44606 Saint-Nazaire, France

Abstract. The efficiency of an optical sensor embedded in a composite structure strongly depends on the interfacial adhesion between the optical fiber coating and the surrounding solid material and on the environment humidity. Moisture diffusion can induce a decrease of the mechanical stiffness and strength of organic matrix composites. The present work reports on the study of the interfacial adhesion of an optical fiber embedded in a composite material. A sample composed of optical fibers embedded in an epoxy vinylester resin or polyester resin with glass fibers was studied to evaluate the influence of the water diffusion and the glass concentration on fiber bonding.

Keywords: Immersion duration · Water diffusion · Optical fiber · Interfacial adhesion · Smart composite materials · Pull-out test

1 Introduction

The development of so-called smart structures, such as composite structures into services in which optical fibers such as sensors have been integrated, has led to characterization studies and analysis of the implementation of these fibers in composite materials that will be used in civil engineering work (bridges, footbridges, gateways, beams, etc.) [1]. The term "smart" comes from the dual functionality provided by the insertion of the optical fiber sensors into composite materials that provide information on the continuous evolution of damaged structures under mechanical stress [2].

But composite materials present some disadvantages, such as susceptibility to moisture absorption, and a non-negligible loss of tensile stiffness and strength can be obtained with immersion in water. For a long immersion period, the effect of water leads to significant damage to the interfacial shear strength [3].

The diffusion of water depends on immersion time and on many other parameters [4] (temperature, composition of resin and curing agent, …) and the damage caused by the effect of water at the interface surface between an optical fiber embedded in the resin/reinforcement fiber composite is more complex to analyse and to quantify.

G. I. Gheorghe (Ed.): ICOMECYME 2020, LNNS 143, pp. 26–33, 2020.
https://doi.org/10.1007/978-3-030-53973-3_4

In this work, we developed composite specimens (resin + glass fibers) (with different fiber glass volumes) in which an optical fiber is embedded. These samples were immersed in distilled water (at laboratory temperature) for different durations (between 6 and 60 days). We will focus on the effect of the immersion duration in water on the optical fiber polymer/composite interface and the change of the rupture force when the optical fiber was submitted to a tensile test. Pull-out tests on optical fibers were carried out to measure the effect of water diffusion and glass concentration on fiber bonding.

2 Samples and Tensile Tests Used

2.1 Sample Preparations

The optical fibers used are silica fibers with one layer of acrylate polymer coating. The diameter of the cladding is 80 μm and the coating diameter is 101.8 μm. These fibers are designed to be used at elevated temperatures and pressures in aggressive chemical environments.

Two commercial resins have been used. The first resin is an ortho-phthalic polyester resin (POLYLITE 420-731) and Methyl Ethyl Ketone Peroxide (MEKP) has been used as a catalyst for initiating the polymerization of this polyester resin. The second polymer resin used in this study is a mixture of epoxy vinylester resin provided by Derakane 470-36, and a catalytic system composed of Styrene, Percadox 16 and Trigonoc C, provided by DC Pultrusion. The glass fibers used for the preparation of the composites come from "Roving Tex". The powdered Perkadox 16 is firstly diluted in the styrene to form the "styrene/Perkadox 16" system. Then the Trigonoc C and the epoxy vinylester resin were added respectively. Before characterizing the interfacial adhesion between the optical fiber and the resin/glass fiber composite material, a simple system composed of an optical fiber embedded in resin was studied. The system is a unidirectional composite, where glass fibers have the same sample length direction and optical fiber direction. The objective was to obtain basic data on a simple system. A typical sample preparation consisted of pouring the resin into a mould and introducing the optical fiber into the middle of the sample. In the case of composites, the impregnated glass fibers were successively superimposed and the optical fiber gently introduced. Special care was taken to ensure that the optical fiber was straight. The set of moulds were then inserted into an oven at 80 °C for 2 h to polymerize the sample.

2.2 Tensile Set-Up Used

The single fiber pull-out test was a direct method for measuring the fiber matrix interfacial properties by the evaluation of the interfacial shear strength. Single pull-out tests were performed on optical fibers. The samples used for pull out tests were carefully prepared. If we applied a gripping force (imposed by the jaws of the LLOYD LR 50K tensile testing set-up) (Fig. 1) to the right hand side of the tested specimen and if the optical fiber was pulled, the fiber breakage occurred at the bit. Thus, a notch was introduced into the specimen and the jaws clamped the right hand part of the specimen. The crosshead speed was set at 1 mm/min, which corresponds to a strain rate of about 0.04 min^{-1}.

The debonding force F_d was taken as the maximum force preceding partial debonding. Different samples obtained with various glass fiber concentrations were tested.

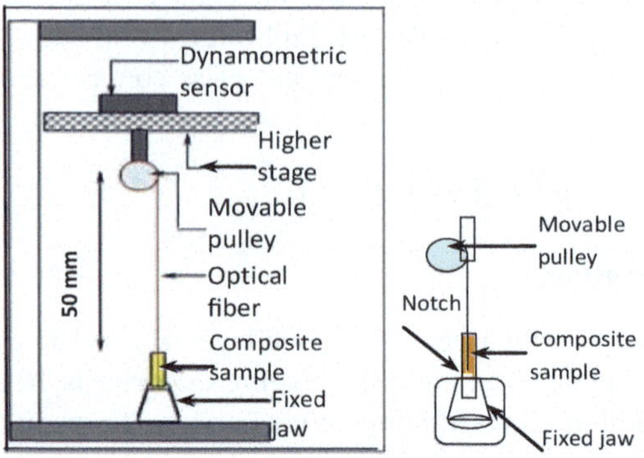

Fig. 1. Description of the dynamic tensile test bench and detail of used composite sample

2.3 Water Effect

The composite we studied will be a component of a civil engineering structure subject to bad weather. During manufacture, this structure receives protective layers to combat water diffusion. However, these protective layers are destroyed during its life and therefore the water diffuses through the composite material.

In recent years, many studies have focused on the analysis of moisture absorption in composite materials [5]. It has been shown that absorbed water can modify the elastic-plastic behaviour of the resin and lead to decohesion of the matrix/fiber interface, and composite performance degradation may occur during use [6].

When composite material was aged in water, water diffusion had a noticeable effect on material properties and a study of the mechanical behaviour change of the interface between the matrix and optical fiber surface was undertaken.

3 Results and Analysis

3.1 Diffusive Behavior

The hygroscopic ageing tests were carried out on resins samples, in order to identify their diffusive behavior. The initial weights of the samples were recorded. Thereafter, specimens were immediately placed into deionized water. The change in mass was measured using a balance with an accuracy of 0.1 mg. The weight gain versus the square root of time ($\sqrt{t}$) curves for the composites and neat resins samples were determined in order to follow their moisture absorption kinetics.

Figure 2 shows the evolution of the moisture uptake as a function of the square root of time, obtained for the two resin samples. We could note that these samples present a Fickian diffusion behavior. The maximum moisture absorption capacity of the neat vinylester sample is twice that of the polyester resin (3% versus 1.5%, respectively). Thus, the maximum moisture absorption capacity of the neat vinylester is at least three times that of the composites specimen.

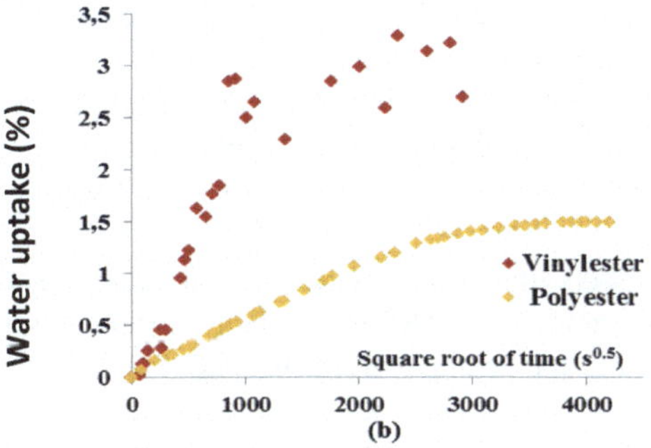

Fig. 2. Moisture absorption curves for two resin samples

The interfacial debonding stress of silica optical fiber/polyester (O.F./polyester) and silica optical fiber/vinylester before hygroscopic aging by using the single-fiber pull out test was studied. Figure 3 shows the force-displacement curves of the O.F./polyester and O.F./vinylester samples obtained via pull out-test before ageing.

The force-displacement curves of the O.F./polyester system exhibit a linear elastic portion up to F = 2 N, followed by a continuous force decrease. In this specific case, the decline of force might be controlled by the friction on the total embedding length of optical fiber (Fig. 3a). On the other hand, the force-displacement curves of the O.F./vinylester system shows a linear elastic region, until a force of 8 N, followed by a discontinuous, brutal decrease of force from 8 N to 0 N. In the following, we will study composites with an optical fiber embedded in vinylester resin.

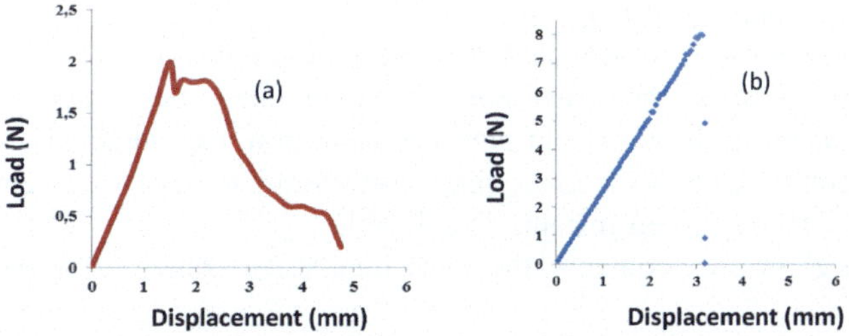

Fig. 3. Force-displacement curves of the unaged O.F./Polyester (a) and O.F./Vinylester (b) samples obtained owing to pull-out tests.

3.2 Change of Water Content Versus Immersion Time

The studied samples were either made of pure vinylester resin in which an optical fiber was embedded or made of resin and glass fibers with an embedded optical fiber. The composite samples contained different volume fractions of glass (0%, 40%, 50%, and 70%). The different samples were 30 mm in length, 10 mm in width and 3 mm in thickness. For each case, four identical samples were used to obtain an average value.

The most used model to explain water recovery in polymers is classically based on the Fick's diffusion law [7] which gives water content changes versus the square root of time. Figure 4 gives the water content change for resin samples and for resin samples with an embedded optical fiber.

During the first 15 days, the water content increased in a linear manner with versus the square root of immersion time. After a transition period extending up to 50 days, the sample was saturated and water weight reached 1.25%.

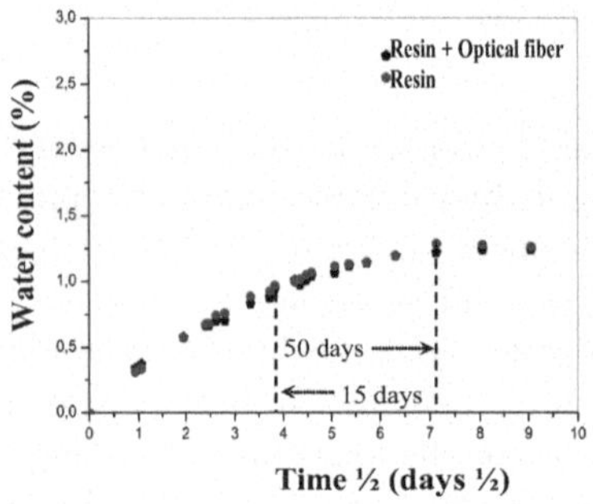

Fig. 4. Water content versus immersion time for vinylester resin samples

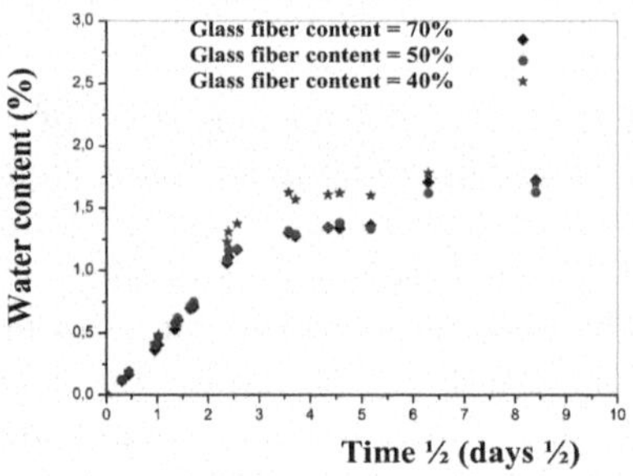

Fig. 5. Water content for different glass fiber volume content for composite with vinylester resin

Composite samples with various glass fiber concentrations were submitted to immersion in distilled water (Fig. 5).

Composite samples with 50% and 70% glass fiber content have similar behaviour. During the first 9 days, the water content had a linear behaviour, followed by a transition period up to 36 days, and then the saturation was obtained.

For the sample with 40% glass fiber concentration, the water content linearly increased up to 10 days then instantly stabilized.

For all the glass fiber contents, the same final water content of 1.7% (Fig. 5) was obtained.

There was less water diffusion in resin samples (Fig. 4) than in composite samples (Fig. 5) where micro-voids and micro-cracks exist.

Figure 6 gives the moisture change for three glass fiber-polyester composite samples containing of volume fraction of fiber respectively equal to 17%, 21% and 22%. According to this figure, the three samples exhibit an almost linear change of moisture uptake for several months until a pseudo plateau indicating that the saturation of the diffusion process is reached.

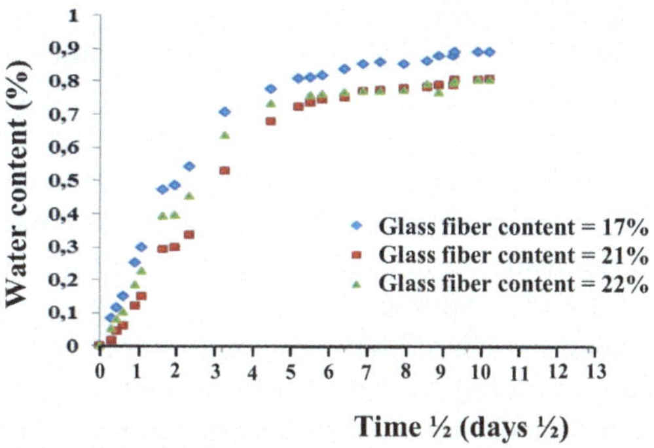

Fig. 6. Water content for different glass fiber volume content for composite with polyester resin

3.3 Tensile Test Analysis

Tensile tests on the optical fiber embedded in samples were undertaken:

a) for vinylester resin samples immersed in distilled water for 10 days, 29 days and 60 days,
b) for composite samples (with 40%, 50%, and 70% glass fiber content) for 60 days.

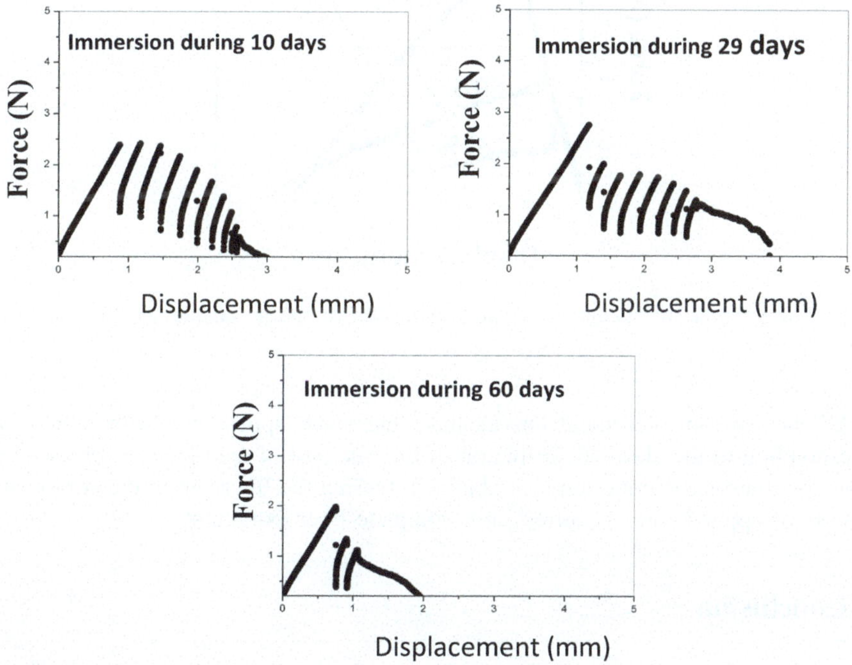

Fig. 7. Force-displacement curves for different immersion durations in water for optical fiber embedded in vinylester resin samples and submitted to pull out test.

**Aging of Vinylester Resin Samples*

Figure 7 gives the force-displacement curves for samples after different water aging durations. All the force-displacement curves have the same appearance. The tensile deformation of the free part of the optical fiber gives the elastic behaviour. After the maximum value of the applied force, a harsh decrease is obtained and is indicative of decohesion of the optical fiber polymer. The second part of the curve indicates the optical fiber sliding in the resin. At first, this slippage is unstable (stick-slip), and then stabilizes at the end of the fiber pulling.

The mean interfacial stress decreased versus aging time. The water diffusion produced tensile radial stresses at the optical fiber/resin interface and led to the interface decohesion. On the other hand, the water diffusion was near saturation after 15 days and the sample was completely saturated after 50 days. A decrease of resin/fiber adhesion was then obtained.

**Aging of Composite Samples*

Figure 8 gives the tensile tests results for composite samples with different glass fiber contents and vinylester resin aged in distilled water for 60 days.

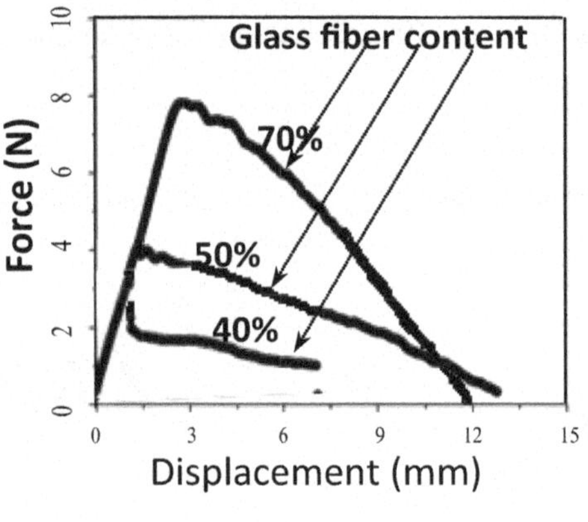

Fig. 8. Force-displacement curves from pull-out test for composite samples aged in water for 60 days

All the force-displacement curves have the same appearance with a first zone corresponding to the elastic deformation of the free part of the optical fiber. After this phase, the interface decohesion was obtained. During the fiber slip in the composite, a decrease of applied force occurred until complete fiber extraction.

4 Conclusion

Using the tensile test procedure, the interfaces resin/optical fiber and composite/optical fibers were characterized after different aging durations in distilled water.

There was less water diffusion in resin samples than in composite samples where micro-voids and micro-cracks exist. The damage of the interface between the polymer coating and the composite surface was due to the chemical-physical action of diffused water.

For vinylester resin/optical fiber samples, water diffusion led to polymer/matrix interface damage from an aging duration of 15 days and force values presented a low decrease up to an aging duration of 60 days. For the case of composite/optical fiber samples, the greater the glass fiber content, the lower the water damage at the polymer interface. We can also mention that glass fibers are stiffer than vinylester resin; therefore this effect can seriously affect the stiffness of the optical fiber surrounding the composite material.

References

1. Karbhari, V.M.: Fiber reinforced composite bridge systems–transition from the laboratory to the field. Compos. Struct. **66**, 5–16 (2004)
2. Yashiro, S., Okabe, T., Takeda, N.: Damage identification in a holed CFRP laminate using a chirped fiber Bragg grating sensor. Compos. Sci. Technol. **67**, 286–295 (2007)
3. Chen, H., Miao, M., Ding, X.: Influence of moisture absorption on the interfacial strength of bamboo/vinylester composites. Compos. Part. A. **40**, 2013–2022 (2009)
4. Joliff, Y., Belec, L., Heman, M.B., Chailan, J.F.: Experimental, analytical and numerical study of water diffusion in unidirectional composite materials – interphase impact. Comput. Mater. Sci. **64**, 141–145 (2012)
5. Karimt, A., Felcher, G.P.: Interdiffusion of polymers at short times. Macromolecules **27**, 6973–6979 (1994)
6. Hanson, T., Glaesemann, G.: Incorporating multi-region crack growth into mechanical reliability predictions for optical fibres. J. Mater. Sci. **32**(20), 5305–5311 (1997)
7. Popineau, S., Rondeau-Mouro, C., Sulpice-Gaillet, C., Shanahan, M.: Free/bound water absorption in an epoxy adhesive. Polymer **46**(24), 10733–10740 (2005)

Predictive Maintenance in Correlation with Industry 4.0 and the Circular Economy

Petrin Drumea and Alexandru-Daniel Marinescu(✉)

INOE 2000 - IHP, 14 Cuţitul de Argint Street, Sector 4,
040557 Bucharest, Romania
marinescu.ihp@fluidas.ro

Abstract. The paper is an attempt by the authors to put together some of the modern concepts of the technical and economic development that are becoming increasingly important especially in the industrial field. In this sense, the predictive maintenance applied in the field of hydraulic drives and its integration in the circular economy and in Industry 4.0 are analyzed first of all, along with reliability and mechatronics. The connecting elements of all these technical-scientific fields are the defect and the behavior in time of the hydraulic installation. The paper presents some of the works performed in the laboratories of the Institute as well as the modern methodologies applied by IHP specialists in the process of the predictive maintenance applying in the hydrostatic drives on the industrial equipments.

Keywords: Predictive maintenance · Industry 4.0 · Circular economy

1 Introduction

This paper analyzes the link between maintenance and reliability and their inclusion in the modern concept of the circular economy. The link between predictive maintenance and Industry 4.0 is also determined, as a support for the development of future technologies. The connecting element of all these technical-scientific fields are the defect and the over time machine behavior. The well-known European 20-20-20 program provides a reduction in gas emissions (CO_2 or equivalent) by 20% compared to 1990, by increasing of energy efficiency and increasing the using action of the renewable resources at least 20% of total energy until 2020. Starting from the European Directive within the Institute have developed non-polluting energy methods and power equipments based on renewable materials and methods have been developed to prolong the life by permanent checking, which does not stop the installations in the activity through which the equipment and hydraulic components such as pumps, valves and hydraulic motors are functionally tested. The checks give indications regarding the situation of wear and the stage of the defects, helping the specialists to intervene at the right time, neither too early which would mean additional costs, nor too late which would mean the occurrence of possible catastrophic defects. The achievement of the proposed goal was made by thermographic methodologies with infrared rays applied in hydraulic drive systems, [1, 2] with vibration analysis methodologies [3] or even by elements of tribological analysis. These tests allowed the specialists to create some databases at the

G. I. Gheorghe (Ed.): ICOMECYME 2020, LNNS 143, pp. 34–43, 2020.
https://doi.org/10.1007/978-3-030-53973-3_5

beginning quite smalls, but with good development prospects. In time, it is hoped to reach the possibility of comparing the catalog thermograms with the operational thermograms obtained on an experimental basis of the installations under construction.

2 The Link Between Maintenance and Reliability, the Circular Economy and Industry 4.0

2.1 General Elements About Maintenance

The maintenance is defined as a combination of all technical, administrative and managerial actions that are taken during a life cycle of an equipment in order to maintain or restore its ability to perform the desired function (according to European Standard EN 13306). In this sense, the maintenance includes many activities of which of interest in this material are the fault detection, repair process, replacement of such elements or subassemblies and the service activity. These activities must be thought of from the design phase, in order to be sure that at the end of the life cycle, as many materials as possible can be reused. The maintenance is an active technical process, which starts from the design phase of a product, by following a few simple rules:

- Achieving of an efficient filtration system, cheap and without negative energy implications,
- Optimal design concerning avoid the energy heat and noise reduction,
- Making of simple schemes with few equipments that influence each other and with few elements that contain non-recyclable materials,
- The use in the realization of the machine of some components with high reliability,
- Establishing of a precise program of preventive maintenance with the possibility of taking over elements from predictive maintenance or even the proactive maintenance, which will increase the lifespan.

2.2 Introduction to Reliability

The reliability is the ability of a product to maintain within the prescribed field parameters since design and specified in the data sheets and catalogs. The reliability prediction is considered in a statistical sense. The environmental and operating conditions include the constraints on the use of the system/product whose reliability we determine, as long as these conditions remain unchanged for all similar products. The event that causes a product to stop working is called a malfunction or failure. To estimate the reliability of systems, the failures are discrete events in time, which depend on the process by which the failure state is reached. From this point of view, failures can be instantaneous or gradual [4]. The specialists must to think, from the design phases, about ways to improve the functional performance of hydraulic installations and specially to keep them in operation for as long time as possible. In the design phase, an analysis of the reliability indicators is usually made, then extending the results of the analysis to the finalization of some concrete elements of conception knowing that the link between reliability and maintenance is the "defect", in the both cases looking for this should happen as late as possible and eventually as predictably as possible. In practice, general measures are taken to improve maintenance and specific measures. The designer will also make an analysis of the materials used and how to reuse them at the end of the life cycle, which should not be the last. As is easy to understand, the basic element of reliability is the defect or loss of usability that can occur suddenly or gradually. A hydraulic product is suitable for use when it meets the basic parameters required by hydraulic systems, such as flow rates, maximum working pressures, the response times, stability, precision performance adjustment in the process of use. Of course, if the hydraulic products are used on mobile machines, the size and weight also start to matter. As already mentioned, these parameters vary over time and as a result it becomes very important the period of operation as long as they are kept within the limits established and accepted from the design phase. The problem of maintaining the quality and the functional capacity of the equipment over time has been a long time coming, but it has remained at the level of appreciation for a long time. Only when the reliability indicators and the measurement methods acquired technical-scientific values could be overcome the phase of assessments such as high reliability, or low or acceptable. In the technique and technology it work both with the operational reliability, the one in operation and with the experimental reliability, the one made in the specific laboratories. In the both cases, the causes of the faults and the time of occurrence of the faults are followed.

2.3 The Circular Economy

The circular economy is an economy model that involves sharing, reusing, repairing, renovating and recycling the existing materials and products as much as possible. In this way, the products life cycle is extended. The basic idea is that we need to waste as little as possible and reuse as much as possible. For this, it seems essential to find ways to program wear, which leads to increase in the lifespane of the products and as a result to a reduction in the consumption of raw materials and matters. All these elements can be found in reliability studies, but also in maintenance studies. Although recently, the idea of circular economy is based in many situations on traditional elements of the use of hydraulics in complex systems and equipments. In this sense, it is found that the rather high prices of hydraulic equipment and systems and their rather expensive maintenance have created elements of the circular economy over time. Unlike the economy of consumer goods in which the linear economic model has been used for years, that after manufacturing and consumption had the stage “throw in the trash” in hydraulic drives were used and still practice principles such as: reuse, repair, maintain, remanufacture, everything to reduce the volume of garbage and transform it into raw material. Since 2015, the European Commission has launched a 54-point a program promoting the idea of changing the economic model, which considered that resources are abundant, cheap and waste-generating that does not matter economically or as a polluting factor, with a new model in which the basic elements are the reuse, repair, renovation and recycling of materials and products, to increase their lifespan and useful life [5].

The transition to the circular economy is not only a vogue and a subject of internal and international discussions but also a source at the disposal of humanity for the development of society on sustainable principles. From the beginning, it must be specified that the products, equipments, machineries must enter into a redesign phase to prevent the giant increase of waste, because the reuse of components and materials bring huge savings.

2.4 Industry 4.0

The term Industry 4.0 was born in Germany and represents in addition to the name of a Government Research Council and the name of a strategic technical project, as well as the name of a research platform. The expression 4.0 means that it would be the initiation of the fourth industrial revolution and that it is based on software and mechatronics products. There are similar initiatives in the world, but with different names. In the USA there is the Industrial Internet Consortium (IIC), in Japan there is the Industrial Value-Chain Initiative (IVI), in France there is the Industry of the Future. A very important role is played by the Industrial Internet of Things (IIOT) which describes the connection of computational power with mechatronic systems and their interconnection via the Internet. In the field of hydraulic equipment manufacturing, the main problem is the manufacturing process, which is mostly done manually and which even in the introduction of equipments is also done manually, which ultimately leads to increased the manufacturing time and facilitation of the errors that influence the quality of the final product. The commissioning process during the third industrial revolution involves limitations in the fields of communication, information, coherence, the existence of a relatively large number of manual activities, as well as too many individual actions that reduce the profitability and the flexibility. Most of these deficiencies are solved within the concept of Industry 4.0. With the transition from manufacturing to Industry 4.0, mechatronic components and cyber-physical systems will move to a symbiosis of human activity with the production process which in own turn is driven by digital interfaces and whose maintenance, supply and logistics are already included in computer programs [6]. The modern industrial equipments that becomes the basis for the transition to Industry 4.0, although are mechanically, hydraulically or electrically operated, in most situations involves special elements such as cyber-physical systems, the Internet of Things (IoT), or the Industrial Internet of Things (IIoT). It is important to mention that since from the conception phase, the specialists, especially those with extensive IT knowledge, ensure that mechano-hydraulic systems can be easily maintained and especially that the reliability indicators will have values that allow an extended lifespan in the conditions of performing the functions at designed and desired parameters. The basis of Industry 4.0 is the digitalization of companies, but this can only be achieved if the employees are prepared for a technical-scientific evolution in this sense. As a result, the maintenance becomes an active factor in the operation of complex machines and equipments, and the modern solutions such as predictive maintenance are both integrated in Industry 4.0 and the circular economy. The industry that encompasses IoT in a number of tens of billions (see IDC) of components at this year's level (2020) is clearly committed to the modernization way. Connecting of the objects to the Internet especially through sensors or other equipments that use teletransmission methods leads to the achievement of intelligent identification and of proper management of things. There is already a lot of equipments for technological lines capable of communicating intelligently. Although traditionally Industry 4.0 is related to production, lately there have been concerns about connecting to the Internet, computers and complex databases and in the field of maintenance, through the emergence of predictive and proactive maintenance, or total productive maintenance (TPM).

3 Predictive Maintenance

Predictive maintenance is the solution for maintaining of hydraulic systems and equipments that allows early detection of defects that are then processed by specific methods. It can be appreciated that this is a new concept that is based on a functional analysis of the equipment or system, after which we will move on to establishing critical areas for which a permanent surveillance method is determined so that the defects are detected from the incipient phase. In recent years and starting especially from the aviation area, the basis of maintenance is the time at which the intervention is made to change some elements, which is established based on previous experience and which is known as preventive maintenance. This methodology often leads to situations in which good equipment is replaced, or even worse it occurs after the failure. As a result of some time, a modern variant is practiced, economically efficient and easy to include in the modern methodologies for the management of complex installations, most of the times hydraulically operated, called predictive maintenance. The final result of predictive maintenance is the improvement of reliability indicators, but especially the increase of the lifespan of hydraulic installations, the reduction of raw materials and of the consumed energy, the reduction of material maintenance costs and consequently the inclusion in a circular economy system. As always in this working situation, an economic-financial analysis must be made to determine in a concrete situation if it is better to intervene when the defect occurs as in the case of the corrective maintenance, if we intervene by changing important components at predetermined deadlines based on previous experiences, such as the case of preventive maintenance, or we replace some components when the tracking sensor systems indicate the appearance and the level at which a certain deterioration of the machine operation reached as in the case of predictive maintenance. Predictive maintenance uses vibration monitoring, or the use of thermal imaging cameras and even tribological determinations, which allows the transition from the time of change of some components based on previous experience, to their replacement depending on the starting of defects occurs. In fact, the predictive maintenance methods allow us to make a timely monitoring of the faults, in addition of making their location as well as, since beginning passed to a precise significant areas selection of the technological system. The predictive maintenance is the solution for maintaining hydraulic systems and equipments that allows early detection of defects that are then processed by specific methods. From what has been shown so far, the following can be distinguished as basic elements that define the two types of maintenance:

- Preventive maintenance includes: lubrication (replacement of lubricating agent), periodic replacement of gaskets; checking the assemblies and them tightening; running of some test sequences through the electronic controller, etc.;

- Predictive maintenance can be materialized by an automatic states monitoring subsystem, i.e. a series of sensors (monitoring bearing temperatures, viscosity and/or clarity of lubricant, the shaft/shafts speeds, load/working torque, stability of translation axis positions temperatures, viscosity and/or clarity of lubricant, the shaft/shafts speeds, load/working torque, stability of translation axis positions, coaxiality, etc.) and a control unit capable to reacting (pumping fresh lubricant, speeds reducing, risk states signaling, logging the history of working conditions, alerting or stopping the installation in case of imminent failure, etc.). The main benefits are related to the productivity increasing by reducing of unforeseen and costly interventions and improving the production process through a predictive planning of interventions and spare parts. Predictive maintenance is actually a strategy that predicts when changing a component or equipment that could fail, but based on received informations from special sensors mounted in the important functional areas. The predictive maintenance is a method used to replace, when it's the case, parts that may have or even have problems with quality or operating capacity. Three requirements must be met to use the monitoring process:

- the damage rate of the product must be carried out in a rhythm that allows the detection of the defect and the specialized intervention time,
- the deterioration process must be well determined, based on precise data, in order to be easily detected, and the changes of parameters to be relevant,
- measuring equipments (mainly the sensors) to be possible to read, analyze and interpret by the existing specialists.

4 Non-invasive Methods of Predictive Maintenance

4.1 Vibration Analysis

Hydraulic equipment and systems have many moving parts and as a result most of the time there are shocks and vibrations that always become a source of damages. These mechanical vibrations result in the appearance of noises not only hard to bear but also with destructive consequences in the system. It is shown that the noises are also caused by the appearance of the cavity caused by the strangulation of the pump suction. Figure 1a and b shows respectively a special noise test stand with and without cavitation of a hydraulic gear pump [3].

Fig. 1. (a) and (b) Noise testing stand of a hydraulic gear pump [3]

4.2 Thermography

One of the basic methods of the non-invasive study of the systems in operation that are the basis of predictive maintenance is the infrared thermography. It is known that maintaining a good operating temperature range of hydraulic drive systems, as well as the viscosity of the working fluid, will lead to increased productivity and decreased energy losses. These goals can be easily achieved with the help of the infrared thermography.

At INOE 2000-IHP Bucharest, within an Innovation Checque, in 2016 this method was promoted and applied in order to assess the degree of wear and functionality of hydraulic pumps and cylinders, by making thermograms followed by a comparative analysis of them [7].

Figure 2a and b show the stands for hydraulic cylinders and Fig. 3a and b show the thermal images corresponding to them.

Fig. 2. (a) and (b) Testing stands for hydraulic cylinders [7]

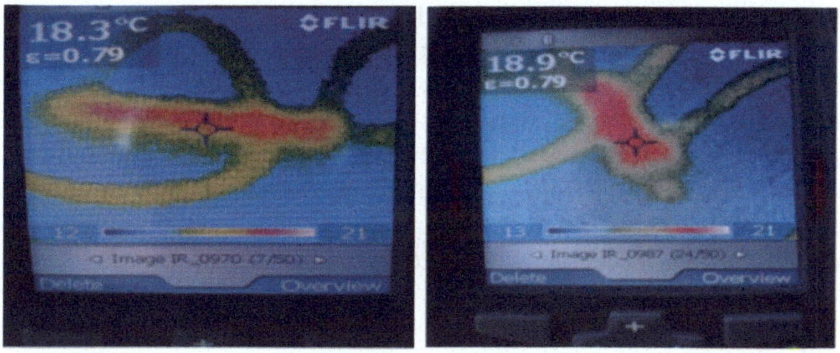

Fig. 3. (a) and (b) Thermal images of the analyzed hydraulic cylinders [7]

Also shown in Fig. 4a and b are a stand for pumps with axial pistons, respectively the overall thermal image resulting from thermography.

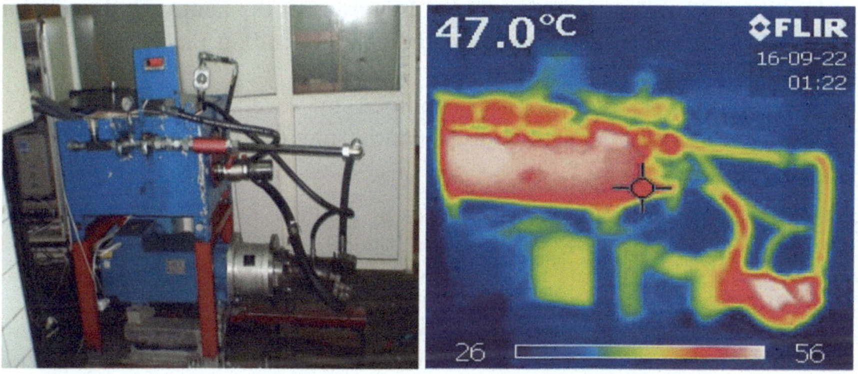

Fig. 4. (a) and (b) Testing stand for axial pumps and the resulting thermal image [1]

4.3 Tribological Analysis

Predictive maintenance is based on the analysis and diagnosis of the conditions that can predict delicate future events. The chosen calculation methodologies will determine the remaining operating times, taking into account that predictive maintenance involves diagnosing defects. In this way, measures can be taken to extend the life of the car. Based on the fact that 70% of hydraulic system failures occur due to the appearance of contaminating particles in the working fluid, it can be said that they cause rapid wear during operation, with serious consequences that sometimes lead to total blockage of the system [8]. Careful and continuous monitoring of the working fluid is therefore required. By continuously monitoring the hydraulic oil used either online or offline, it is possible to know if it is still alive and especially for how long.

By operating the system with degraded or contaminated oil, the life of the entire system is reduced. That is why the degradation of the oil and the penetration of water into the working fluid must be seriously checked. This technology, close to the proactive method, can provide accurate informations about a possible malfunction of the machine and can thus improve the reliability of the equipment. An oil analysis program takes samples of oil from the installation and processes them to determine its functional condition. These results are compared with the limits accepted from the beginning and the right decisions which are required are made.

5 Conclusions

5.1 Maintenance, in all known variants is a science, with a high level and multiple applications.

5.2 Modern methods of maintenance, preventive, predictive, proactive, total productive fit perfectly into the modern fields of the circular economy and Industry 4.0.

5.3 Maintenance in all variants is coupled by means of the fault with the reliability, the both activities serving to increase the time of good operation of the hydraulic system.

5.4 Predictive maintenance has been imposed by several non-invasive working methods such as vibration analysis, thermography or tribology.

Acknowledgements. This paper has been funded by the Romanian Ministry of Research and Innovation under Programme I-Development of national R&D system, Subprogramme 1.2–Institutional performance–Projects financing excellence in R&D&I, Financial Agreement no. 19PFE/17.10.2018, Phase 4, while the scientific results presented were obtained under the National Research Programme NUCLEU, Financial agreement no. 18N/08.02.2019, Project acronym: OPTRONICA VI, Research theme no. 2, titled: "Advanced research on developing synergic border architectures used in solving global challenges and improving knowledge-based competitiveness", Phase titled: 'Research on increasing the reliability of hydrotronic products and systems by non-invasive predictive and total productive maintenance (TPM) methods'.

References

1. Marinescu, A.D., Popescu, T.C., Enache, L., Safta, C.A.: Researches on specific malfunctions diagnosis of hydraulic drive systems equipments using the infrared thermography method. In: Proceedings of the 22nd International Conference Hervex, Băile Govora, Romania, pp. 218–224 (2016)
2. Marinescu, A.D., Cristescu, C., Popescu, T.C., Safta, C.A.: Assessing the opportunity to use the infrared thermography method for predictive maintenance of Hydrostatic Pumps. In: Proceedings of the 8th CIEM International Conference on Energy and Environment, Bucharest, Romania, pp. 270–274 (2017)
3. Marinescu, A.D., Orăşanu, N., Safta, C.A.: Vibroacoustic predictive investigations on normal or defective operation of hydrostatic pumps. In: Proceedings of the 25th International Conference Hervex, Băile Govora, Romania, pp. 169–176 (2019)

4. Safta, C.A., Marinescu, A.D., Cristescu, C., Popescu, T.C.: Reliability modelling of hydrostatic equipments and system drives. In: Proceedings of the 24th International Conference Hervex, Băile Govora, Romania, pp. 157–164 (2018)
5. Drumea, P., Popescu, A.M., Dumitrescu, C.: Fluid power industry 4.0 and circular economy. In: Proceedings of the 25th International Conference Hervex, Băile Govora, Romania, pp. 191–197 (2019)
6. Raphael, A., Murenhoff, H., Schmitz, K.: A survey of Industry 4.0. in the field of fluid power – challenges and opportunities by the example of field device integration. In: Proceedings of the 11th International Fluid Power Conference, Aachen, Germany, pp. 15–25 (2018)
7. MAGUFTER Project. http://www.ihp.ro/magufter/index.htm. Accessed 08 May 2020
8. Drumea, P., Dumitrescu, I.C., Hristea, A., Chiriță, C.: Methods of diagnosing malfunctions in hydraulic actuations. In: Proceedings of the 22nd International Conference Hervex, Băile Govora, Romania, pp. 212–217 (2016)

Aspects Regarding the Modelling of Geometric and Strength Calculations of Worm Gears Using CAD Applications

Aurel Mihail Ţîţu[1,2(✉)] and Alina Bianca Pop[3]

[1] Lucian Blaga University of Sibiu, 10, Victoriei Street, Sibiu, Romania
mihail.titu@ulbsibiu.ro
[2] The Academy of Romanian Scientists, 54, Splaiul Independenţei, Sector 5, Bucharest, Romania
[3] SC TECHNOCAD SA, 72, Vasile Alecsandri Street, Baia Mare, Romania
bianca.bontiu@gmail.com

Abstract. The new processes of plastic deformation of the gears eliminate some of the shortcomings of the old methods of their manufacture, while reducing the production costs and the time of production in series. This scientific paper addresses the issue of worm gears. The focus has been on geometric and strength calculations for this type of gear, using different CAD applications. The use of these applications substantially reduces the time before the actual development of the processing activities, as well as the early encounter of possible problems.

Keywords: Worm gear · Gears · CAD applications

1 Introduction

Among current mechanical transmissions, gear transmissions have the widest use, ensuring compact and reliable constructions for the entire power range of the machines (from a few watts to tens of thousands of kilowatts) [1–3].

The subject analysed in this scientific paper, refers to the worm gears. This topic has been analysed by many authors, including [4–6].

The worm used in the worm gear can be machined; by turning, milling and grinding. Regardless of its shape (evolutionary, archimedical or convoluted), it can be processed by turning on ordinary lathes, using knives with straight-edged edges, taking into account the way of generating the respective surface.

Disc milling can be applied to any type of worm, the profile of the cutter must have the shape of the profile in the normal section on the worm propeller. In the case of convoluted worm, the profile is rectilinear.

The worm finishing operation is required due to the fact that they are always subjected to a heat treatment. The finishing operation can be done by sanding or grinding.

G. I. Gheorghe (Ed.): ICOMECYME 2020, LNNS 143, pp. 44–57, 2020.
https://doi.org/10.1007/978-3-030-53973-3_6

Grinding is performed, with abrasive powder and oil, on the lathe, with the help of wooden pliers. This operation cleans the surface of the thread of the snail of oxides resulting from hardening, but can not ensure an improvement of the thread profile [3, 7].

Worm grinding involves machines corresponding to the type of worm.

2 Worm Wheel Processing. Basic Aspects and Perspectives

Small worm wheels can be machined on the universal milling machine, with a milling cutter - disc - module, the division being performed with the divider head, and the feed, in the radial direction.

In order to obtain a more correct shape of the teeth, the diameter of the disc cutter must be as close as possible to the size of the auger diameter. Often the worm is built to the diameter of the available disc cutter [8].

A substantial improvement of the gear of the worm wheel is obtained using a milling cutter - the worm-module corresponding to the worm, with the help of which the worm wheel is finished, on the universal milling machine, the worm wheel rotating freely between the tips.

For series production, worm wheels are machined on milling machines with screw milling cutter, by three methods: radial, tangential and combined [9].

An economical method of processing worm wheels on gear milling machines, in the case of single or a small number of wheels, for which the production of a worm cutter would not be economical, is to use instead of the worm cutter knives properly mounted on a mandrel [10, 11].

The finishing of the worm wheels is usually done on the same machines on which their teeth were processed. The tools used are special screw milling cutters, with or without adjustable calliper teeth. Snake-like snails are also used to scrape the fine teeth. In general, however, the finishing operation presents difficulties in that the tool with which the finishing is done must have the smallest possible deviations from the shape and dimensions of the auger with which it will engage the auger wheel and be arranged at the same distance and in the same position as in the aggregate in which the worm gear will work [12].

The new processes by plastic deformation of the gears eliminate some of the shortcomings of the old methods of their manufacture, reducing at the same time the production costs and the time of realization of the production in series.

3 Geometric and Strength Calculation Possibilities for Worm Gears Using CAD Applications

To carry out this study, the AutoDesk INVENTOR PROFFESIONAL application was chosen.

Using the information provided by http://www.girard-transmissions.com, the 3D model of a worm gearbox in the form of a stp file was imported into AutoDesk INVENTOR (Fig. 1).

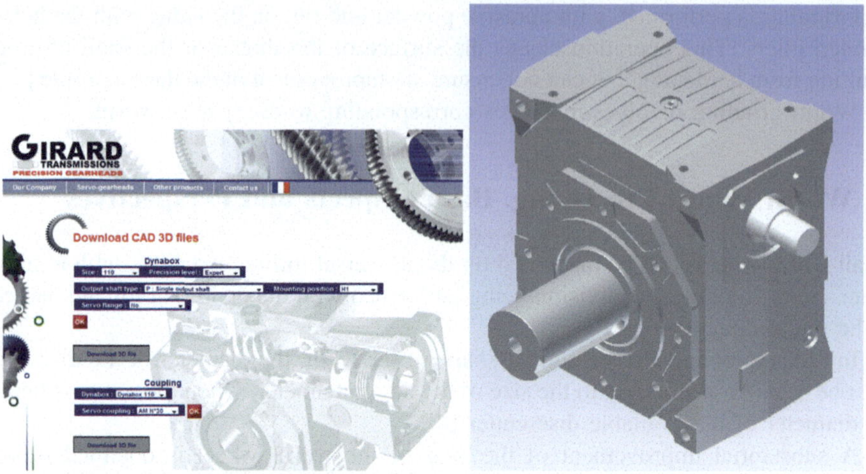

Fig. 1. Import the worm gearbox stp file.

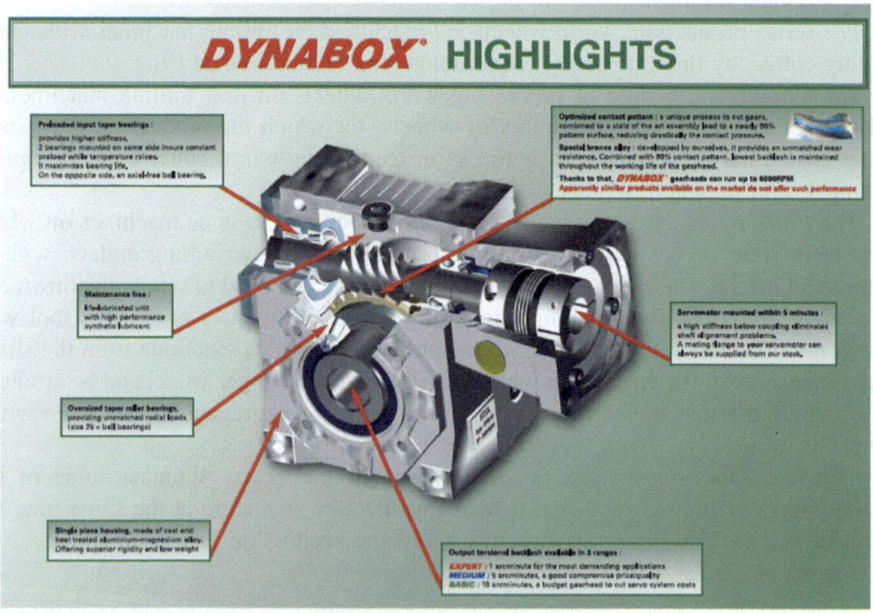

N1		6000		4000			3000			2000			1000									
	i	Torque S5	η	Torque S1	Torque S5	η	Torque S1	Torque S5	η	Torque S1	Torque S5	η	Torque S1	Torque S5	η	E-stop	C1f	ig	Et	Reversibility class	Fr	Fa
DYNABOX 110	5.2:1	567	96	390	666	95	458	779	95	561	937	94	760	1239	92	2289	2	1,85 X 10^{-3}	120	1	21500	16000
	7.25:1	579	95	417	680	95	488	795	95	599	976	94	802	1307	92	2289	2	1,3 X 10^{-3}	120	1	21500	16000
	10.25:1	650	95	449	786	94	522	878	93	638	1047	92	827	1323	90	2289	2	8,5 X 10^{-4}	120	1	21500	16000
	14.5:1	630	93	450	720	92	519	830	91	630	1014	90	810	1247	87	2289	2	6,3 X 10^{-4}	120	1	21500	16000
	19.5:1	670	92	510	815	91	589	943	90	705	1121	88	893	1349	85	2289	1	4,6 X 10^{-4}	120	1	21500	16000
	30:1	790	88	597	955	87	688	1100	85	812	1299	83	1015	1512	79	2289	1	3,5 X 10^{-4}	120	2	21500	16000
	45:1	776	85	583	915	82	665	1037	80	765	1168	78	947	1411	73	2152	1	3,3 X 10^{-4}	120	3	21500	16000
	60:1	683	81	522	815	79	588	905	77	669	1030	73	826	1239	68	2094	0,8	3 X 10^{-4}	120	3	21500	16000
	90:1	645	75	497	765	72	557	847	70	625	944	66	778	1128	60	1941	0,8	1,7 X 10^{-4}	120	3	21500	16000

Fig. 2. DYNABOX catalogue - technical specifications.

Due to the fact that this model does not contain the geometric parameters of the worm gear but only the axial distance (110 mm), using the Design Accelerator module of the INVENTOR application in the following, it is presented how the gear geometry was created using technical data provided by the DYNABOX catalogue and the calculation of the worm gear strength (worm - worm wheel) (Fig. 2).

Using the Accelerator Design module of the INVENTOR application, the steps necessary to make a worm gear and its strength calculation are:

- Gear geometry design (screw and worm wheel design);
- Verification calculation (verification of the resulting data);
- Choice of worm material and worm wheel.

4 Specific Aspects Regarding the Design of the Geometry of a Gear. Proposing Verification Calculations

The first step is to select the Spur Gears tool from which the Calculation option will be selected, in which (Figs. 3, 4 and 5):

- the standard used for the resistance calculation will be specified, in this case ANSI;
- the type of loads to be calculated is specified;
- the type of resistance calculation is chosen.

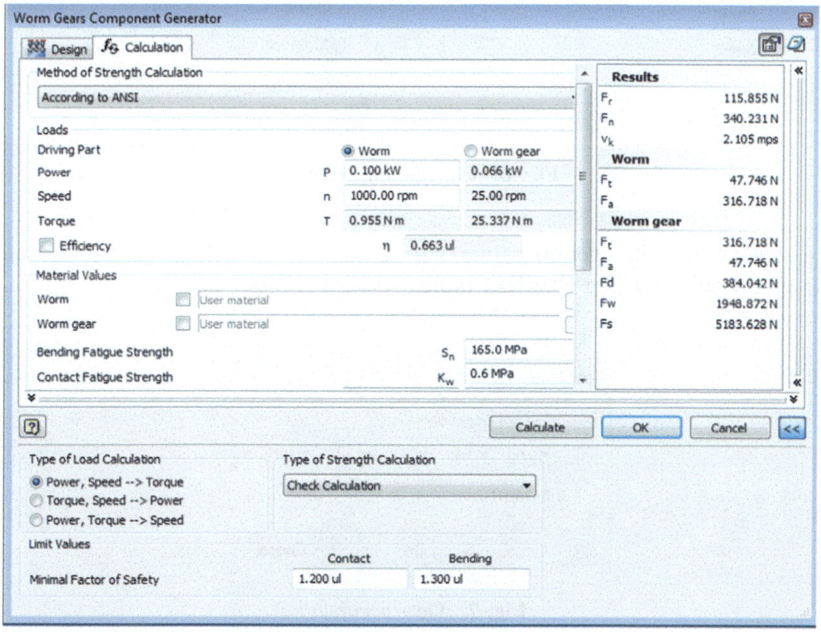

Fig. 3. Choice of standard used.

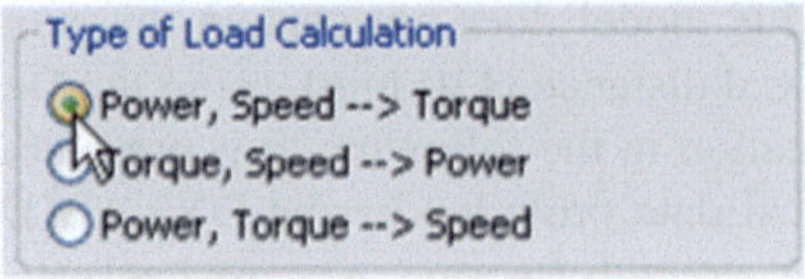

Fig. 4. Choosing the type of loads.

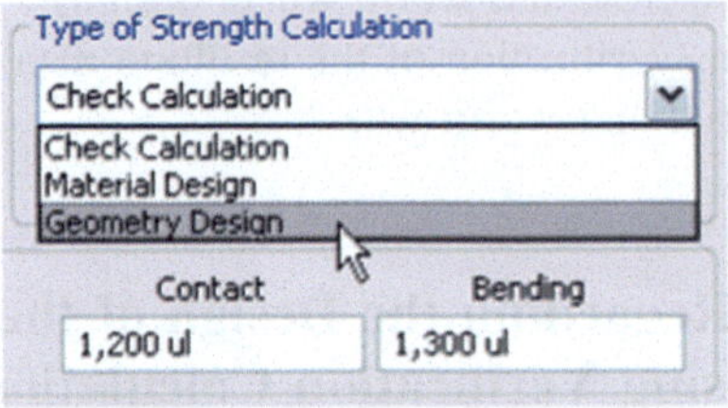

Fig. 5. Choosing the type of resistance calculation.

As an example of calculation we will have as input data the power P = 0.100 kw and a speed n = 1000 rpm (Fig. 6).

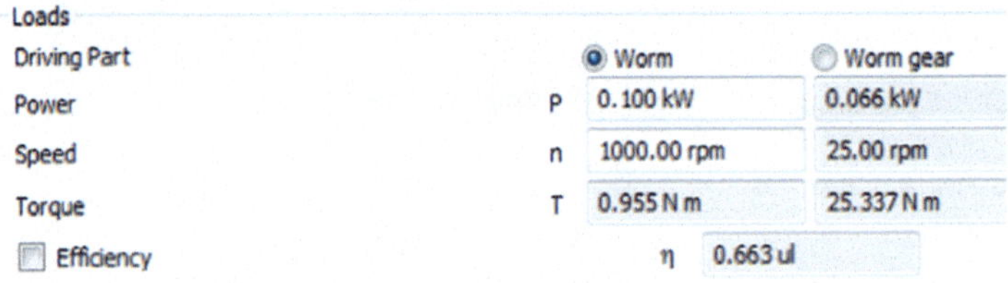

Fig. 6. Calculation example.

We consider that the operating time of the gear Lh = 10000 operating hours and we use the precision according to ANSI indicated next (Figs. 7 and 8).

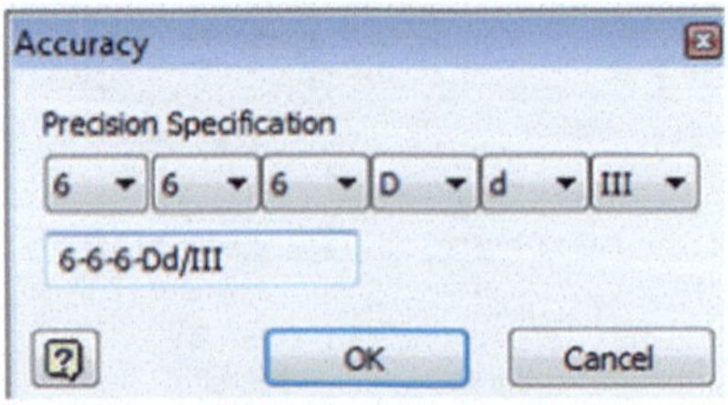

Fig. 7. Gear accuracy.

Fig. 8. Gear life.

In the window corresponding to the factors that influence the operation, we choose as use factor ka = 1.35 (Fig. 9).

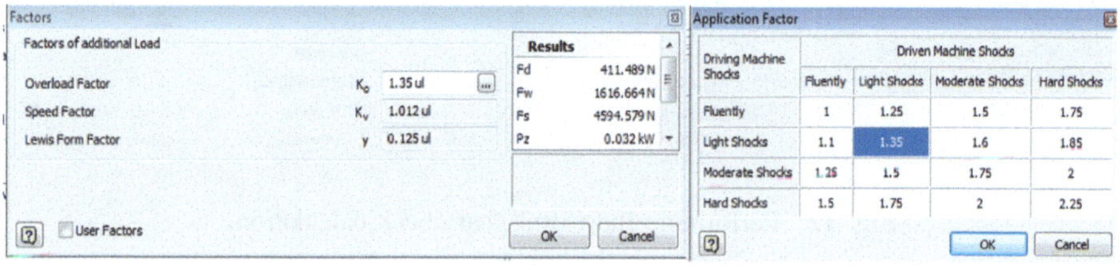

Fig. 9. Selecting the factors of additional load.

Switching to the Design option of the Worm Gear component we choose the value of the desired transmission ratio (in this case 39), the module (in this case 4.5) and the pressure angle (20°) (Fig. 10).

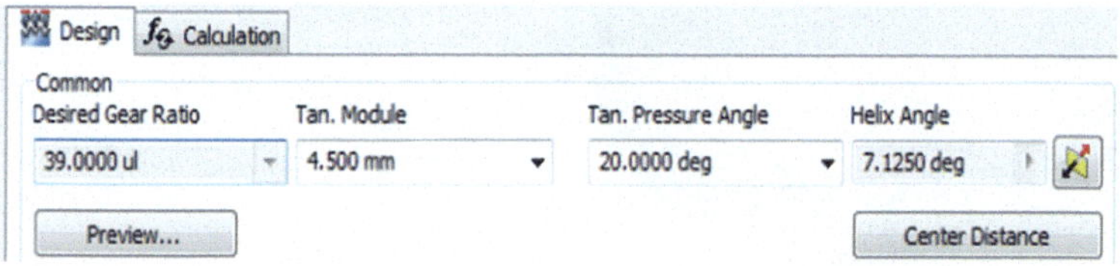

Fig. 10. Choice of transmission ratio.

After performing the above and using the calculation algorithm according to the standard, by clicking on Calculate, the application displays the following results (Fig. 11):

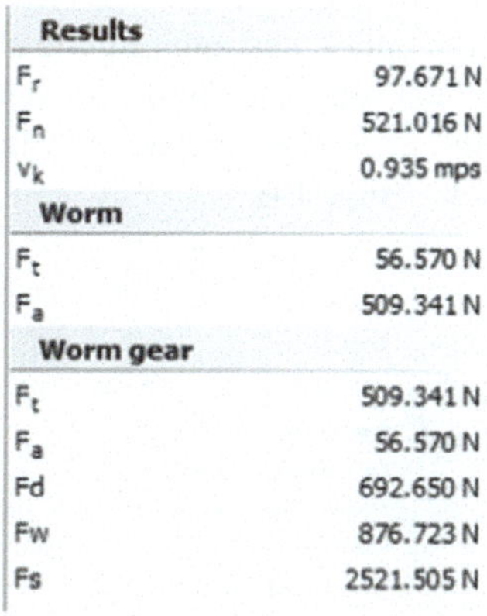

Fig. 11. Displaying results.

By changing the transmission ratio, modulus, power and speed we can get other results.

The first step in performing this step is to select the Check Calculation option.

The length of the auger is considered to be 44 mm and the width of the auger wheel is 17.64 mm (Fig. 12).

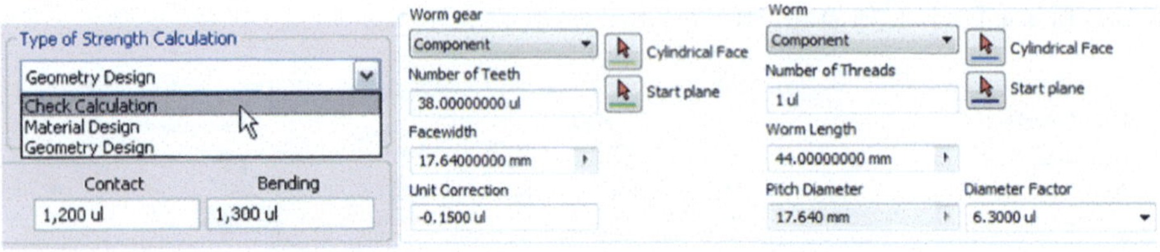

Fig. 12. Performing the worm gear check calculation.

After performing these steps by selecting the Calculate button we check the results obtained (Fig. 13).

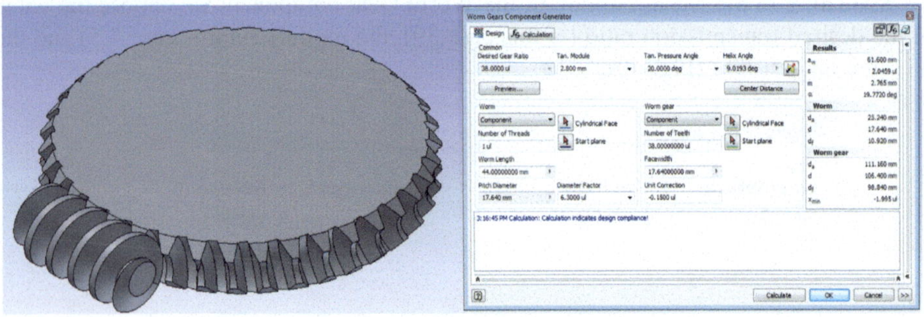

Fig. 13. Checking the results.

5 Materials Used for the Analysed Gears

By choosing the Material Design option and entering the characteristics of the materials of the auger and the worm wheel, the calculation for bending fatigue and the calculation for contact fatigue of the auger and the worm wheel are made, displaying the limits of the two values S_n şi k_w (Fig. 14).

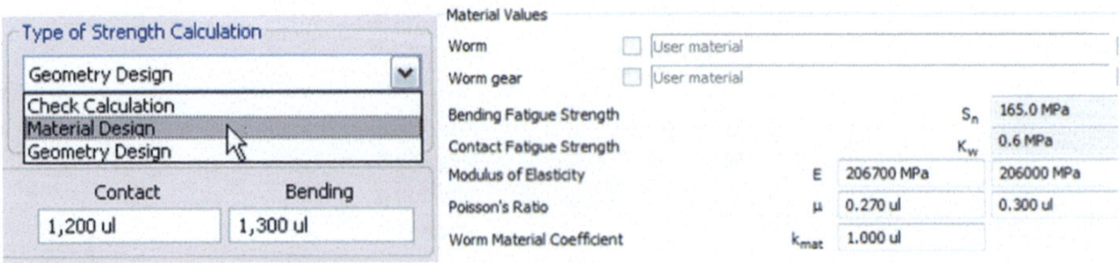

Fig. 14. Characteristics of the worm and worm wheel material.

As an example of calculating the two requests we will apply different materials for the auger and the auger wheel, and to achieve this we must select the Check Calculation option and define the material (Fig. 15).

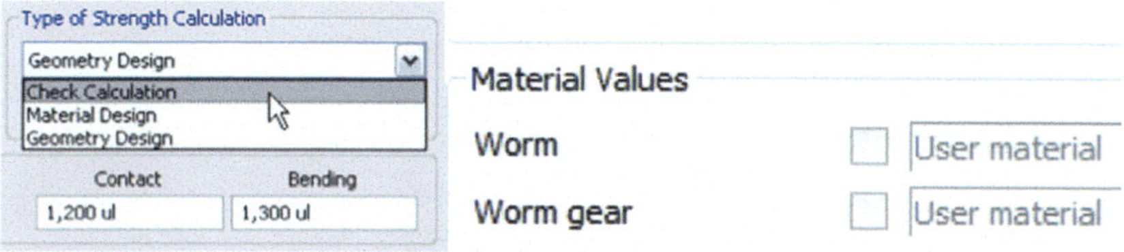

Fig. 15. Defining materials.

The materials are defined using the materials from the ISO standard of the INVENTOR application.

In the present example, the materials indicated below will be used, chosen according to the desired maximum bending and contact fatigue strengths (Fig. 16).

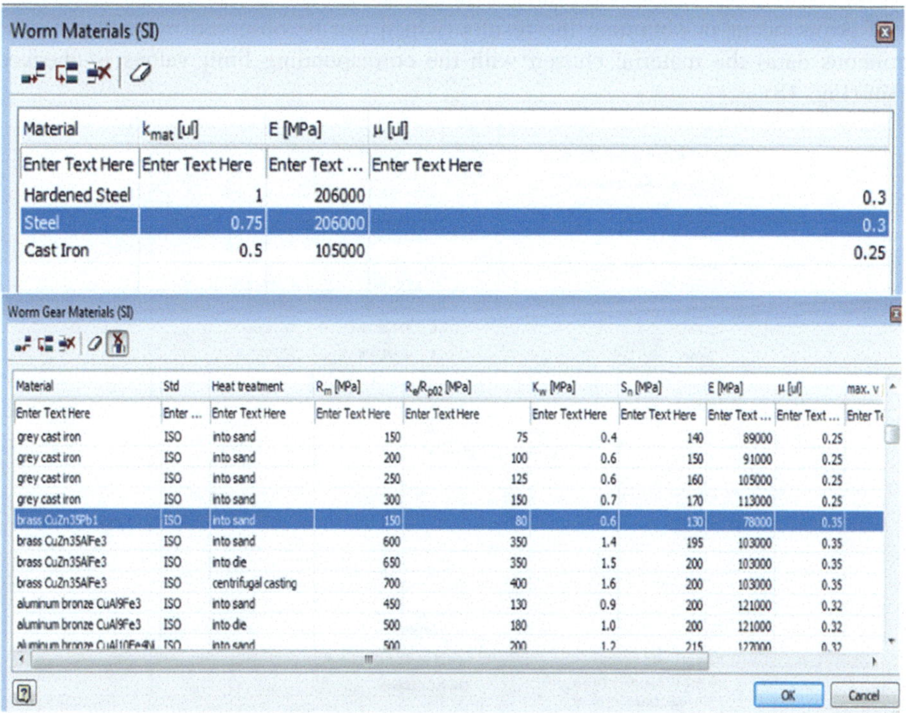

Fig. 16. Choice of materials for snail and worm wheel.

After performing these steps by selecting the Calculate button we check the results obtained (Fig. 17).

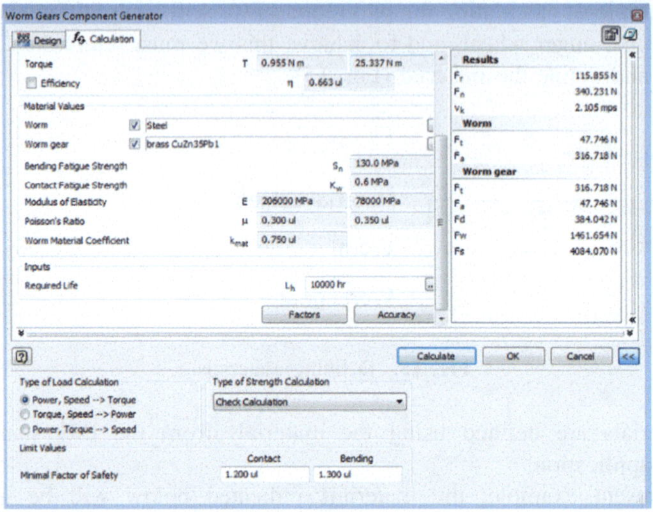

Fig. 17. Rechecking the results.

If errors occur in obtaining the results (which can be observed by reddening the erroneous data) the material chosen with the corresponding limit values is checked again (Fig. 18).

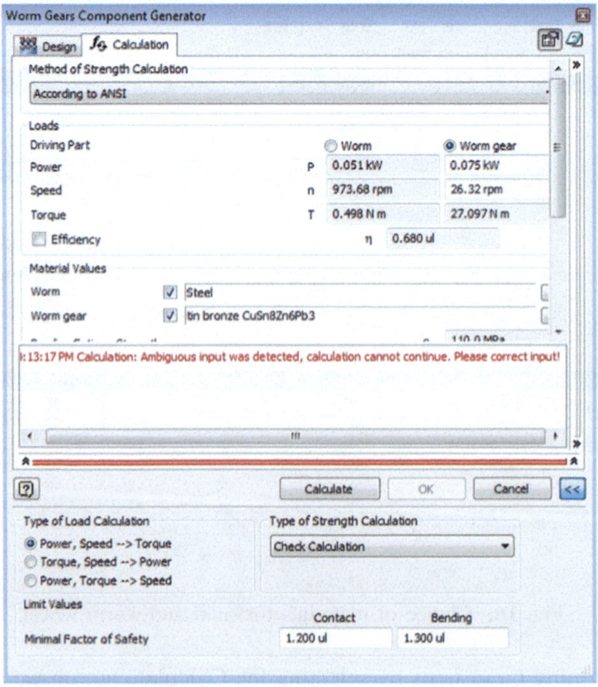

Fig. 18. Indication of errors.

6 Final Conclusions

Finally, the report with the results of the geometric and strength calculations of the worm gear is presented (Fig. 19).

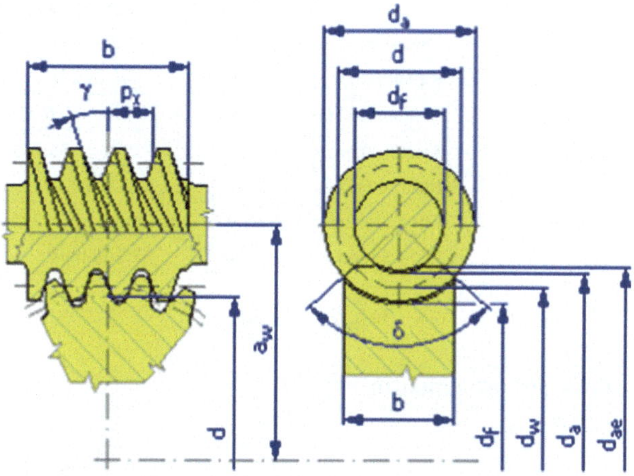

Fig. 19. Worm and worm wheel parameters.

The values of the common parameters are presented in Fig. 20.

Gear Ratio	i	43.0000 ul
Module	m	3.980 mm
Axial Module	m_x	4.000 mm
Helix Angle	γ	5.7106 deg
Pressure Angle	α	19.9086 deg
Worm Diameter Factor	q	10.0000 ul
Center Distance	a_w	110.000 mm
Axis Circular Pitch	p_x	12.5664 mm
Circular Pitch	p_n	12.5040 mm
Base Circular Pitch	p_b	11.809 mm
Lead	pz	12.566 mm
Worm Length	b_1	60.000 mm
Worm gear Width	b_2	29.300 mm
Axial Pressure Angle	α_x	20.0000 deg
Base Helix Angle	β_b	5.3683 deg
Contact Ratio	ε	2.7986 ul
Transverse Contact Ratio	ε_α	2.5654 ul
Overlap Ratio	ε_β	0.2332 ul
Limit Deviation of Shaft Angle	F_β	0.0090 mm
Guaranteed Backlash	j_{nmin}	0.054 mm
Limit Deviation of Center Distance	f_a	0.032 mm

Fig. 20. Common parameters values.

In Fig. 21 are presented the loads values.

		Worm	Worm gear
Power	P	0.100 kW	0.066 kW
Speed	n	1000.00 rpm	23.26 rpm
Torque	T	0.955 N m	27.238 N m
Efficiency	η	0.663 ul	
Radial Force	F_r	-96.546 N	
Tangential Force	F_t	-39.789 N	-316.718 N
Axial Force	F_a	-316.718 N	-39.789 N
Normal Force	F_n	333.108 N	
Circumferential Speed	v	2.094 mps	0.209 mps
Slide Speed	v_k	2.105 mps	

Fig. 21. Loads.

The gears values are presented in Fig. 22.

		Worm	Worm gear
Type of model		Component	Component
Number of Threads	N	1.000 ul	
Number of Teeth	N		43.000 ul
Unit Correction	x	0.0000 ul	1.0000 ul
Pitch Diameter	d	40.000 mm	172.000 mm
Outside Diameter	d_a	48.000 mm	188.000 mm
Root Diameter	d_f	30.400 mm	170.400 mm
Outside Diameter	dae		192.000 mm
Base Circle Diameter	d_b	37.588 mm	161.627 mm
Work Pitch Diameter	d_w	48.000 mm	172.000 mm
Worm gear Chamfer Angle	δ		37.62 deg
Addendum	a^*	1.0000 ul	1.0000 ul
Clearance	c^*	0.2000 ul	0.2000 ul
Root Fillet	r_f^*	0.3000 ul	0.3000 ul
Tooth Thickness	s	6.252 mm	9.149 mm
Axial Tooth Thickness	s_x	6.283 mm	9.195 mm
Limit Circumferential Run-out	F_r	0.0130 mm	0.0400 mm
Limit Deviation of Axial Pitch	f_{pt}	0.0090 mm	0.0140 mm
Limit Deviation of Basic Pitch	f_{pb}	0.0085 mm	0.0130 mm
Virtual Number of Teeth	N_v		43.647 ul
Min. Recommended Correction	x_{min}		-2.314 ul

Fig. 22. Gears.

In the Fig. 23 are presented the material values.

		Worm	Worm gear
		User material	User material
Ultimate Tensile Strength	S_u		500 MPa
Yield Strength	S_y		180 MPa
Modulus of Elasticity	E	206700 MPa	206000 MPa
Poisson's Ratio	μ	0.270 ul	0.300 ul
Bending Fatigue Strength	S_n		165.0 MPa
Contact Fatigue Strength	K_w		0.6 MPa
Bending Fatigue Limit	σ_{Flim}		170.0 MPa
Contact Fatigue Limit	σ_{Hlim}		200.0 MPa
Hardness in Tooth Side	VHV		110 ul
Base Number of Load Cycles in Bending	N_{Flim}		250000000 ul
Base Number of Load Cycles in Contact	N_{Hlim}		250000000 ul
Wöhler Curve Exponent for Bending	q_F		9.000 ul
Wöhler Curve Exponent for Contact	q_H		8.000 ul
Max. Slide Speed	v_{max}		12.000 mps

Fig. 23. Material.

In the next figure, are presented the strength calculations. So, firstly, in the Fig. 24 are highlighted the factors of additional load, and then in Fig. 25 are presented the results.

Overload Factor	K_o	1.200 ul
Dynamic Factor	K_v	1.010 ul
Lewis Form Factor	y	0.125 ul

Fig. 24. Factors of additional loads.

Dynamic Load	Fd	-384.042 N
Surface-fatigue Limiting Load	Fw	3069.230 N
Bending-fatigue Limiting Load	Fs	7594.015 N
Lost Power	Pz	0.037 kW
Max. Dissipated Heat	Q	0.546 kW
Check Calculation		Positive

Fig. 25. Results.

Finally the worm gearbox looks as follows (Fig. 26):

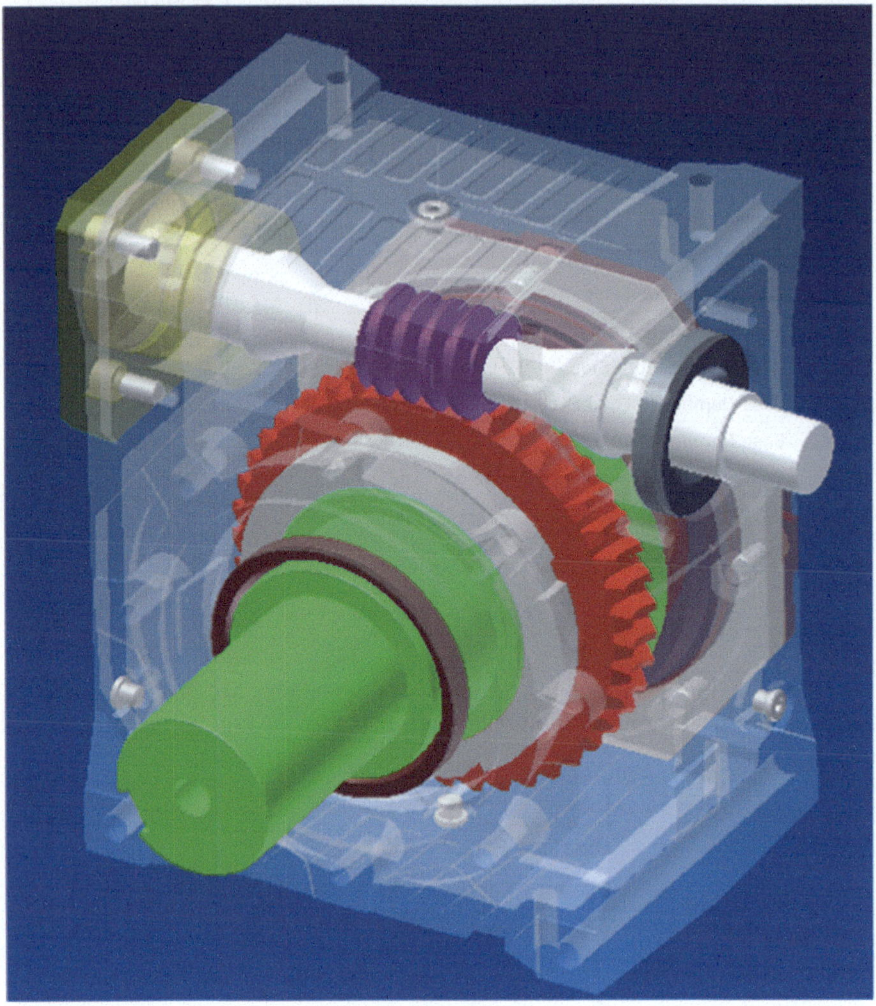

Fig. 26. The worm gear.

The new processes by plastic deformation of the gears eliminate some of the shortcomings of the old methods of their manufacture, while reducing the production costs and the time of production in series.

In this paper we presented a series of aspects regarding worm gears pointing, types of worm gears, processing of worms and worm wheels.

This study presents the elements needed to design a worm gearbox. In the elaboration of the paper, the authors adopted notations in accordance with the current standards.

The most important part of the work is the geometric and strength calculation of the worm gear using the AutoDESK INVENTOR application.

By addressing the main topic of this scientific paper in this way, the time required to perform the calculations presented is significantly reduced.

In the end of this study we can conclude that the calculation indicate design compliance.

References

1. http://www.girard-transmissions.com/downloads.php. Accessed 10 May 2020
2. Kudreavtev, V.N., Gierzaves, I.A., Glukharev, E.G.: Design and Calculus of Gearboxes. Mashinostroenie Publishing, Sankt Petersburg (1971)
3. Chat, T.: Some problems of kinematics calculation of transmission mechanics system. In: Proceedings of the National Conference on Engineering Mechanics, Hanoi, Vietnamese, vol. 2, pp. 7–12 (1993)
4. Cam, N.T.H., Pi, V.N., Tuan, N.K., Hung, L.X., Thao, T.T.P.: Determining optimal partial transmission ratios of mechanical driven systems using a V-belt drive and a helical reducer with second-step double gear-sets. In: Fujita, H., et al. (eds.) ICERA 2018. LNNS, vol. 63, pp. 261–269 (2019). https://doi.org/10.1007/978-3-030-04792-4_35
5. Pi, V.N.: A new study on optimal calculation of partial transmission ratios of two-step helical gearboxes. In: 2nd WSEAS International Conference on Computer Engineering and Applications, CEA 2008, Acapulco, Mexico, 25–27 January, pp. 162–165 (2008)
6. Nguyen, K.T., Vu, N.P., Nguyen, T.H.C., Tran, T.P.T., Ho, K.T., Le, X.H., Hoang, T.T.: Determining optimal gear ratios of a two-stage helical reducer for getting minimal acreage of cross section. In: MATEC Web of Conferences, vol. 213, p. 01008 (2018)
7. Chat, T., Van Uyen, L.: Design and Calculation of Mechanical Transmissions Systems, vol. 1. Educational Republishing House, Hanoi (2007)
8. Pi, V.N., Thao, T.T.P., Tuan, D.A.: Optimum determination of partial transmission ratios of mechanical driven systems using a chain drive and two-step helical gearbox. J. Environ. Sci. Eng. B **6**, 80 (2017)
9. Milou, G., Dobre, G., Visa, F., Vitila, H.: Optimal design of two step gear units, regarding the main parameters. VDI Berichte No. 1230, p. 227 (1996)
10. Hung, L.X., Pi, V.N., Du, N.V.: Optimal calculation of partial transmission ratios of fourstep helical gearboxes with second and fourth-step double gear-sets for minimal mass of gears. In: The International Symposium on Mechanical Engineering, ISME, Ho Chi Minh City, Vietnam, September 2009, pp. 21–23 (2009)
11. Magyar, B., Sauer, B.: Calculation of the efficiency of worm gear drives. Power Transm. Eng. **9**(4), 52–56 (2015)
12. Deng, J.: Introduction to Grey system theory. J. Grey Syst. **1**(1), 1–24 (1989)

Control of Drive Motors for Humanoid Robot Head

Tudor Catalin Apostolescu[1], Ioana Udrea[2], Georgeta Ionascu[2], Silviu Petrache[2(✉)], Laurentiu Adrian Cartal[2], and Lucian Bogatu[2]

[1] Faculty of Informatics, Titu Maiorescu University, Bucharest, Romania
[2] Faculty of Mechanical Engineering and Mechatronics, POLITEHNICA University of Bucharest, Bucharest, Romania
silviupsf@yahoo.com

Abstract. The humanoid robot must mimic the verbal formulation of answers to a set of previously formulated questions. The sound emission of the answer is made with an ordinary rendering system, the sequence of syllables in the acoustic message must be correlated in time with the movements of the lips and mandible, but also with the movements of the head vertically and horizontally, as well as those of the eyes and eyelids, in order to create the impression of dialogue with a human personality. In this paper, the mechanical construction of the robotic head, modeled in CATIA software, as well as control of drive motors for the movements which are performed by the robot's mechanisms, are described. The motors used in the construction of the humanoid head are of two types, direct current motors provided with built-in reducer and encoder for use in highly precise positioning and speed, and stepper motors whose position control and their bringing to the initial position, considered as a zero position, can only be performed by commanding a certain number of steps. The control of all motors is performed by means of a data acquisition board, which, in turn, receives the commands formed by the LabVIEW program. In this program, specific files have been developed for the set of movements that the robot must execute.

Keywords: Human robot head · Motion control · Drive motor

1 Introduction

The design of multimodal robot-type interfaces with humanoid appearance must have the following aesthetic characteristics:

- an attractive appearance, so that humans can naturally establish a relationship with it;
- the robot must have a natural and intuitive interface (relative to inputs and outputs), so that a human can interact with it using natural communication channels. This allows the robot to receive and perceive human signals;
- the robot must have sufficient sensory, motor and computational resources for real-time activities performed in the dynamic interaction with humans.

G. I. Gheorghe (Ed.): ICOMECYME 2020, LNNS 143, pp. 58–71, 2020.
https://doi.org/10.1007/978-3-030-53973-3_7

An iconographic physiognomy consisting of two eyes with eyebrows and a mouth is universally accepted and can portray a simple range of emotions needed in interaction with people. The use of emotions and human face expressions in the context of human-robot interaction enjoys increased attention. Made for research or general use purposes, these robots can be included in two categories: interactive robots with a certain degree of anthropomorphism and humanoid robots. The latter are equipped with a head provided with possibilities for movement and intelligent interaction with the environment.

A structure with a high degree of anthropomorphism, the Aryan robot head [1] presents physiognomy elements with 8 independent movements, i.e. 8 degrees of freedom: neck oscillation, neck tilt, mandible, left eye rotation, right eye rotation, eye oscillation, eyebrows raising and arching. Each of them is made with a specific mechanism actuated with a servomotor, the type of motor being chosen according to the control circuit. The face of the Aryan robot can express various states such as calm, amazement, anger and fear.

At NASA laboratories was made the so-called Robonaut [2], which has two eyes and a neck with two degrees of freedom, which offers the possibility of moving the head up and down, as well as on the left - right side. The eyes are equipped with cameras. The neck mechanism is controlled in real time. The kinematics is based on a set of rotating elements connected in series. A first rotation is performed around the robot column, and another around a horizontal axis. The robot is equipped with eyes endowed with large video cameras for focusing and iris adjustments, as well as a small video camera for the perception of the lateral field. This robot does not aim to establish relationships through physiognomy.

The robot head mounted on RATO [3] is a cheap, simple, easy to integrate in the robot's control architecture and very expressive head. Servomotors, as used in radio controlled cars or air planes, actuate the different elements like eyes, eyebrows, eyelids, mouth, etc. These kinds of motors were chosen because it is quite simple to work with them and they do not need to move heavy parts of the head.

A social robot for daily life activities has been developed in [4]. Its distinctive feature is the use of significantly fewer actuators. Only three servomotors for facial expressions and five for the rest of the head motions have been used. The modular design makes it possible to generate more expressions without addition or modification of components.

Design and control of such robots need complex knowledge regarding the mechanisms, dynamics and intelligent control [5, 6].

In this paper, the mechanical construction of the robotic head, modeled in CATIA software, as well as control of drive motors for the movements which are performed by the robot's mechanisms, are described. The motors used in the construction of the humanoid head are of two types, direct current motors provided with built-in reducer and encoder, and stepper motors. The control of all motors is performed by means of a data acquisition board, which receives the commands provided by the LabVIEW program.

2 Construction of the Robot Head Mechanical Model

The mechanical model performs eight degrees of freedom as follows:

- two degrees of freedom of eye movements;
- a degree of freedom for eyelid movement;
- a degree of freedom of movement of the jaw, which also makes the movement of the lower part of the lips (mouth);
- a degree of freedom making the movement of the upper part of the lips (mouth);
- a degree of freedom for the lateral movement of the lips (mouth);
- a degree of freedom for the rotational movement of the head about a horizontal axis;
- a degree of freedom for the rotational movement of the head about a vertical axis.

The movement of the mouth is made with three degrees of freedom allowing to obtain its opening both vertically and horizontally, thus being able to model the pronunciation.

The CAD environment provided by CATIA software was used for efficient design. The robot head assembly and its prototype are given in [7].

In Fig. 1 details are given regarding the rotation of the eyes (rear view).

For the rotation around the horizontal axis, materialized by the spindle 9, the frame 5 was built. The spindle is bearing in the fixed corner 10. The eyes 6 and 11 are articulated to the frame by the spindles 4 and 12 (so that they can also have the rotational movement around the vertical axis).

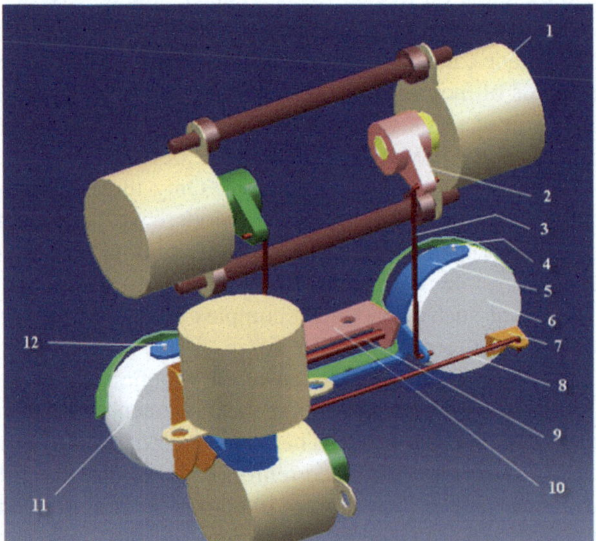

Fig. 1. Details regarding the rotation of the eyes (rear view).

The rotation of the frame around the spindle 9 is made with the quadrilateral mechanism containing the connecting rod 3 and the crank 2 integral with the output shaft of the motor gear 1.

The figure also shows how to synchronize the rotational movement of the two eyes around the vertical axis by using the connecting rod 8, articulated on the corners 7, fixed on the back (chamfered) part of the eyes.

Figure 2 is a rear-left-bottom view that allows detailing how to get rotation around the vertical axis of the eyes. On the back of the eye 5 is fixed the piece 3 in the shape of "C" as a backstage for the mechanism. The crank 2, integral with the output shaft of the gear motor 7, ends with a spherical portion, which drives the slide.

Also, in Fig. 2 it can be seen how to perform the rotation of the eyelids 4. These elements together form a frame which is articulated, like the eye frame, on the same spindle (position 9 in Fig. 1).

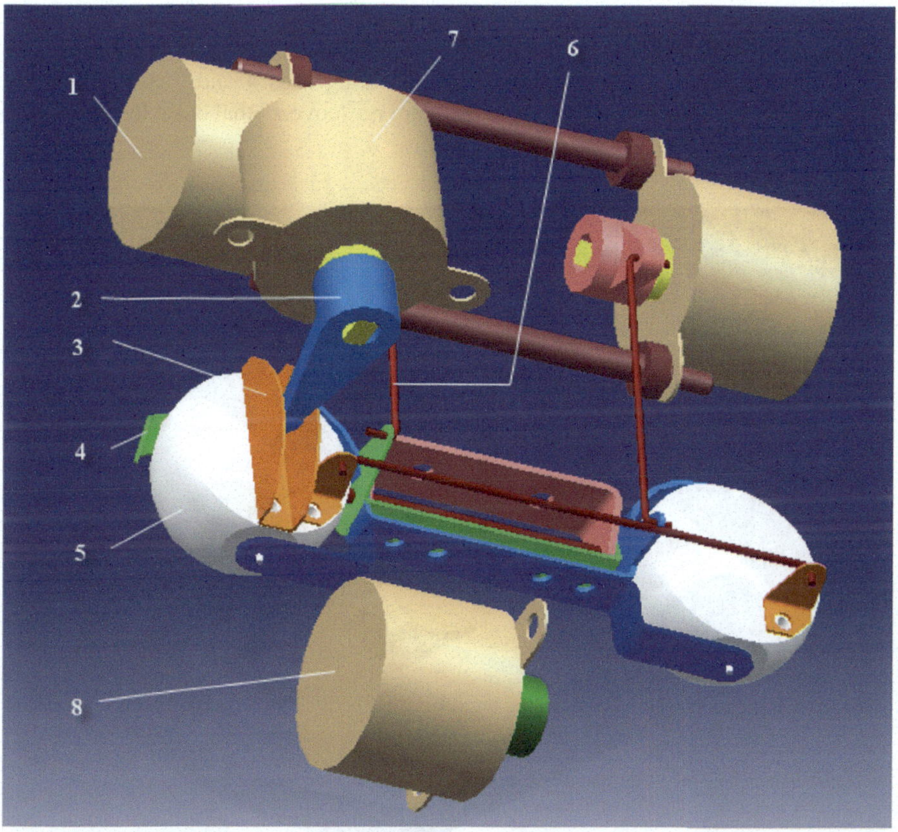

Fig. 2. Details regarding the rotation around the vertical axis of the eyes and the rotation of the eyelids.

The rotation of the eyelid frame is also made with a quadrilateral mechanism, with the connecting rod 6 articulated on the crank of the gear motor 1 (the crank is not seen in Fig. 2. It can also see the motor gear 8 which serves to move the upper part of the mouth; details are given further.

Details regarding the movement of the lips (mouth) are given in Fig. 3 The upper plate and eye movement mechanisms described above have been removed.

The movement of the mandible 6 is made directly from the motor gear 18, its output shaft being coupled to the mandible.

For the up and down movement of the upper part of the mouth 13, the lever 14 is inserted, articulated by the spindle 11 on the bracket 12. The movement comes from the motor gear 17 on whose output shaft the crank 16 is fixed. The transmission of the movement to the lever 14 is made with the connecting rod 15 (articulated quadrilateral mechanism).

An articulated spatial quadrilateral mechanism is introduced for the lateral movement of the corners of the mouth. The levers 7 and 10, whose ends are attached to the corners of the mouth, are articulated on the spindles 9. The movement comes from the motor gear 1 fixed with the corner 2. The crank 3 is fixed on its output shaft. The connection to the levers is made with the connecting rods 4 and 5, using spherical couplings (to simplify the representation there are no "covers" that close the couplings).

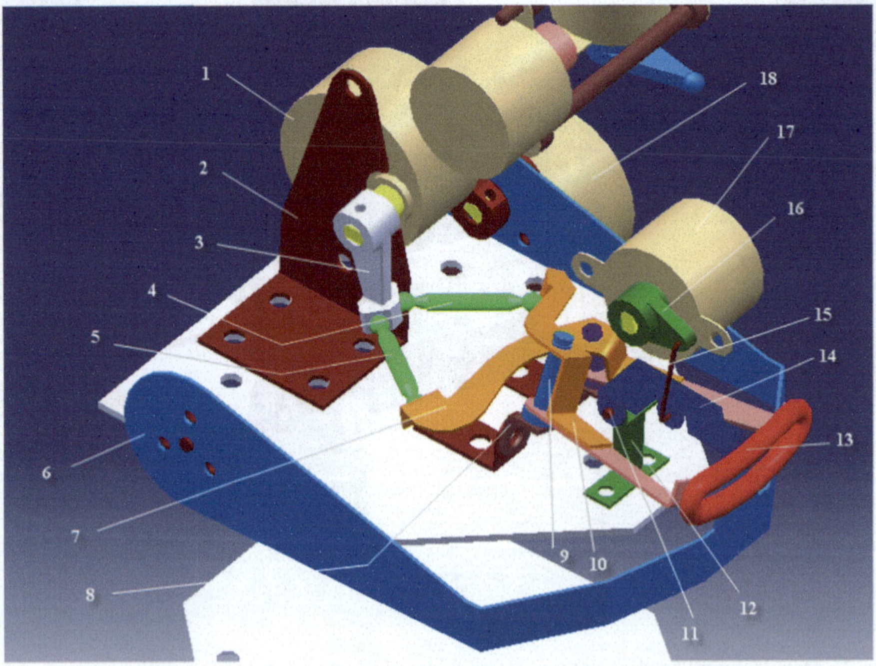

Fig. 3. Details regarding the movement of the lips (mouth).

Such a spatial mechanism was necessary because it was intended that the planes of movement of the corners of the mouth be inclined by 10…20° from the plane of the plate to give a “cheerful” physiognomy to the wide opening of the mouth. The inclination of the spindles 9 is made adjustable, they are being fixed on the corners 8, which can be fixed in the inclined position.

Obtaining the two movements of the neck can be seen in Fig. 4. The figure is a bottom view of the low part of the model. The rotation around the horizontal axis of the neck is made starting from the gear motor 6, whose output shaft is connected, by using the flange 5, to the plate of the lowest part of the construction of the head 4.

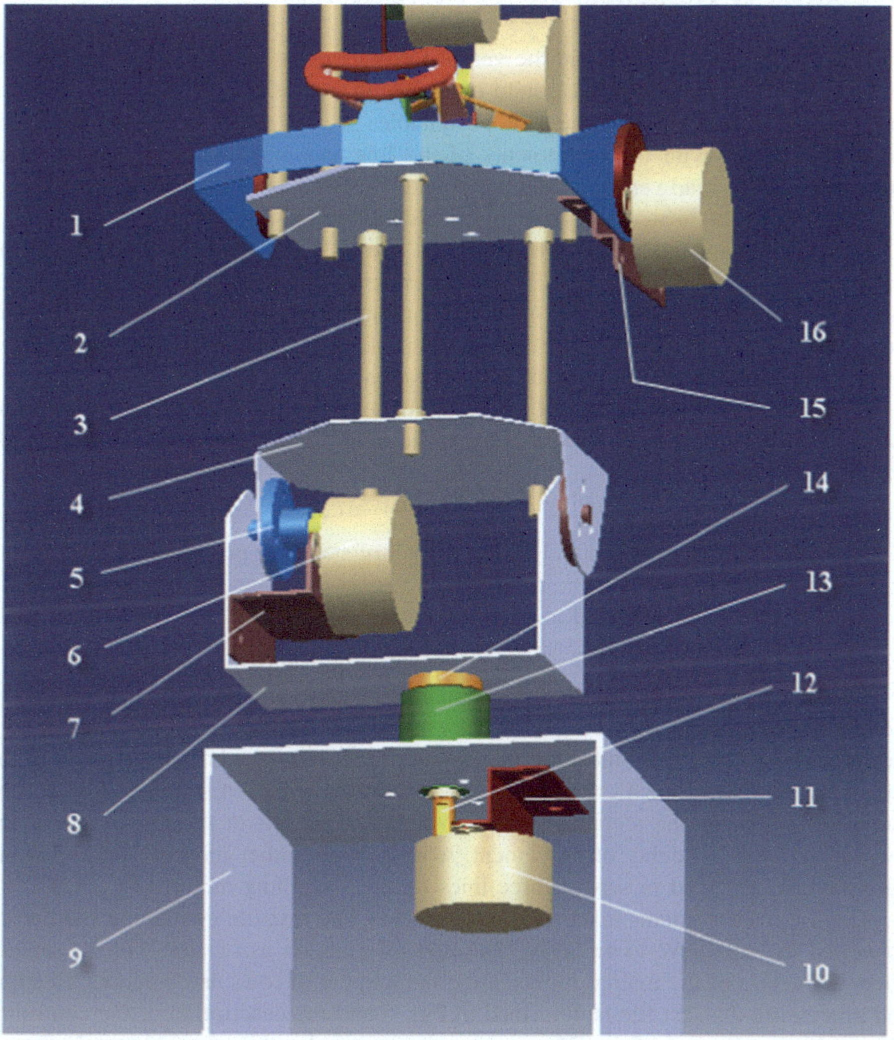

Fig. 4. Details regarding the two movements of the neck.

The fixing between the plates 2 and 4 is made with the tie rods 3. The motor gear is fixed on a piece 8, in the shape of “C”, by means of the corner 7.

The rotation around the vertical axis is obtained starting from the gear motor 10 that acts through its output shaft, the shaft 14 integral with the element 8. The shaft bearing is made by means of 2 radial bearings in the housing 13. The motor gear is fixed to the base 9 of the model, through the corner 11.

The figure also shows a detail of the actuation of the mandible 1. A distinction is made between the gear motor 16 and its fixing bracket 15 on the intermediate plate 2.

3 Control of Humanoid Robot Drive Motors

The humanoid robot must mimic the verbal formulation of answers to a set of previously formulated questions. The sound emission of the answer is made with an ordinary rendering system, the sequence of syllables in the acoustic message must be correlated in time with the movements of the lips and mandible, but also with the movements of the head vertically and horizontally, as well as with those of the eyes and eyelids, in order to create the impression of dialogue with a human personality.

All these movements are performed by the robot’s mechanisms, by controlling the drive motors. Stepping motors (MPP) must perform a certain number of steps, corresponding to the angle of rotation of the motor shaft, and DC (direct current) motors must perform an angle of rotation of the output shaft of the gearbox, which is controlled by the encoder in the motor gear structure, Maxon type.

For each answer that the robot has to simulate, there is information stored in the computer in the form of a text file. Here is the motor control data via the LabVIEW program.

The text file has the following organization:

- the number of columns is equal to the number of states (or positions) of the DC and stepping motors;
- the number of lines is equal to that of the driven motors.

Within a line, the data corresponding to each motor must be entered:

Line 1 - position of the motor shaft 1; Line 2 - position of the motor shaft 2; Line 3 - the relative common delay of the two motors; Lines 4, 5 and 6 - the command data of the stepping motors that control the movement of the neck in a horizontal plane to the left and to the right; Line 4 - direction of rotation, right or left, of the neck; Line 5 - the number of steps of the stepping motor, which is reflected in the amplitude of the movement; Line 6 - delay of movement, relative to the other motors. The following lines are also arranged in sets of three lines for each stepping motor.

The file itself contains a set of control blocks. The text file containing the data is converted in LabVIEW to an array. The first information associated with the matrix refers to the size of the matrix, representing the number of rows and columns. Reading and processing this .txt file is described in Fig. 5 and the description of the control of the stepper motors and of the DC motors, are presented in Fig. 6, respectively Fig. 7.

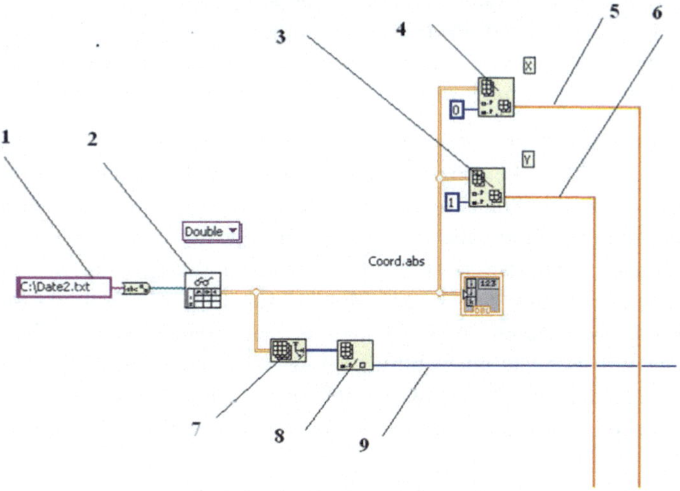

Fig. 5. Reading the data in the .txt file.

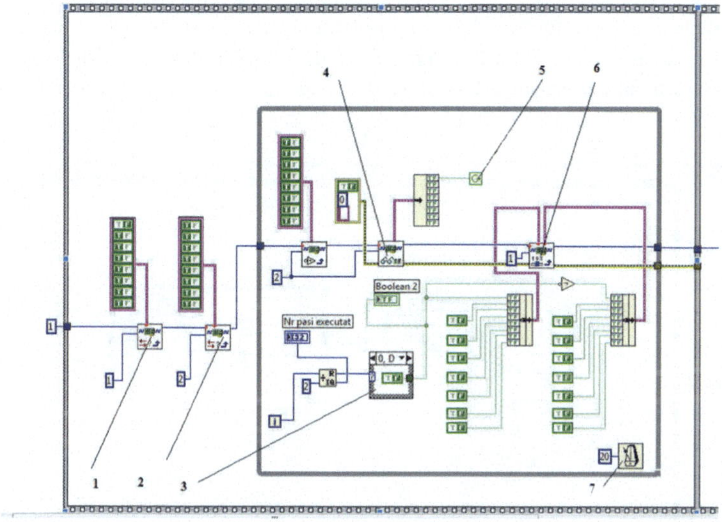

Fig. 6. MPP control with STEP module.

Motion programming is made in LabView NI Motion. The steps (sequences) that are followed to perform the movements are:

1. Creating the database;
2. Reading the database;
3. Axis 1 command;
4. Establishing the reference position for axis 1;

5. Resetting the axis 1 encoder to the corresponding value;
6. Establishing the reference position for axis 2;
7. With the block “Array size” are read the dimensions of the matrix;
8. “Index Array” - the number of rows (MPP positions) that are sent (9) to the MPP.

The control of the stepping motor is done with the help of the I/O part of the data acquisition board 7344. The program part for reaching the desired position and stopping when the stroke limiter is actuated is shown in Fig. 6. Pulses are generated by repeatedly dividing by 2 the value of the meter of a “while” cycle. If the counter value is even (remainder 0 at division) “True” is generated at decision block 3. The output of this block is entered in the “Set I/O MOMO.flx” function, which allows bit 0 of port 1 of the board to be entered, with “True/False” alternative values, thus generating impulses. The stop is made when bit 0 of port 2 has the value “True”, i.e. when the stroke limiter is actuated. Then, the “while” cycle is stopped (cycle stop function 5). The rotation speed is given by the delay block 7. The pulse generation frequency is about 1/0.02 = 50 Hz.

The operation of all stepping motors is made similarly.

The control of DC motors is performed with an individual block. The local variable that generates the execution of a while loop is extracted from the data matrix of the text file. The first For cycle controls the two DC motors provided for the execution of the lips and mandible movement. The For loop must run a number of times equal to the number of columns in the data file. This is made by executing the For loop as long as the counter indicates a number less than the number of columns in the data file.

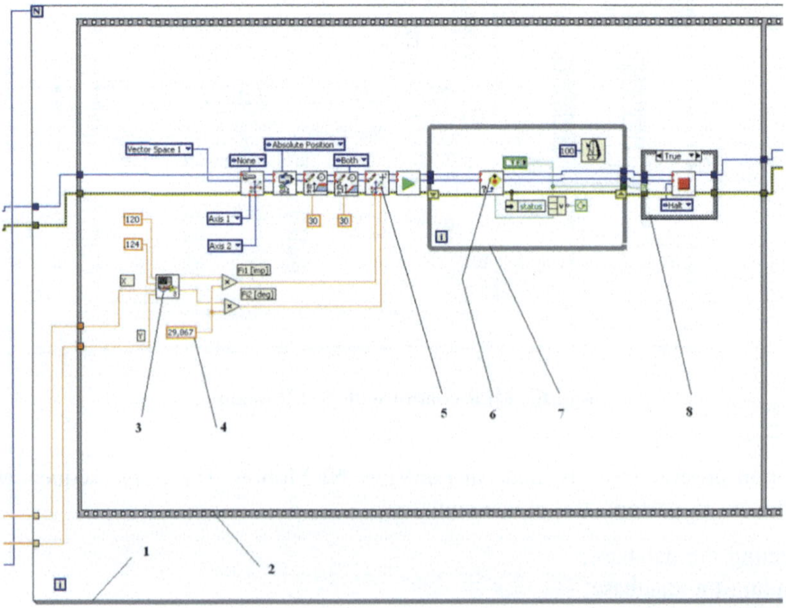

Fig. 7. DC motors control.

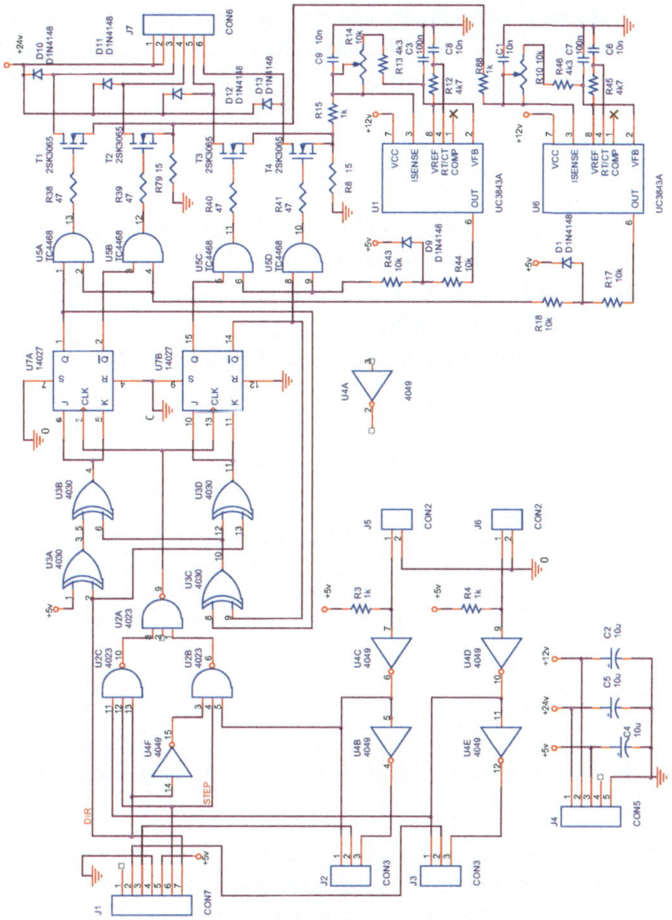

Fig. 8. Electronic circuit diagram of STEP module.

The first DC motor drives the mechanism that moves the corners of the lips to the side. For this, the first data string corresponding to index 1 is extracted from the data matrix.

In Fig. 7, is extracted from the For loop (pos. 1) the part that expresses the displacement at the extracted point from the .txt file (sequence 2). The coordinates of the axis are entered in subroutine 3. The outputs are multiplied by the constant (pos. 4):

$$\frac{128 \cdot 84}{360} = 29{,}867\ \frac{\text{imp}}{\text{deg}} \tag{1}$$

In the above relation were replaced: the transmission ratio of the reducer 84, the equivalent number of pulses given by the encoder for a rotation of the motor shaft: 4 * 32 = 128. The values of the angles Fi1 and Fi2 expressed in pulses at the encoder are entered in the block 5 ("Load Vector Space Position.flx"). A while cycle is next, in which the stop condition is periodically tested with the "Check Move Complete Status. flx" block. If the output is True, the decision block 8 stops the movement.

Similarly, the operation of the second DC motor is controlled, with the role of executing the lifting and lowering movement of the lever that materializes the mandible. For this the data will be extracted from the second line of the data matrix.

STEP Module

The electronic circuit diagram of the STEP module is shown in Fig. 8.

The DIR (sense) and STEP (step) control signals are input signals for the counter made with the JK bistable circuits (U7) and EXCLUSIVE OR (U3). The transition diagram of the states can be seen in Fig. 9.

Stepping motors are controlled in "unipolar, full step" mode.

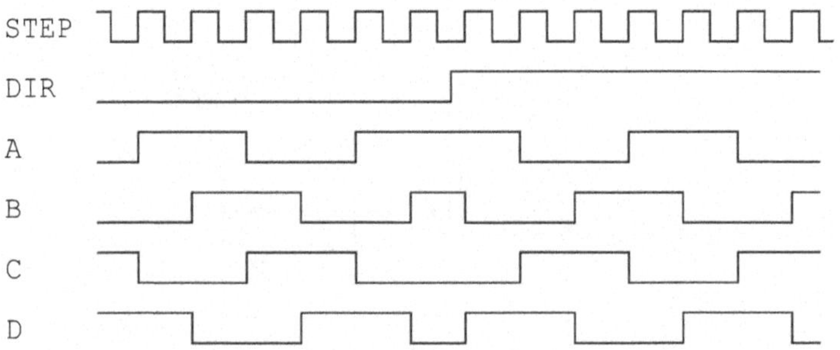

Fig. 9. Transition diagram of the states at the stepping motor control.

The circuit made with the logic gates U2 and U4-E ensures the protection of the mechanism by locking the controls to the stepping motor, in the sense that when a stroke limit signal occurs, the motor control is not allowed only in the opposite direction of the stroke end.

NMOS transistors are used to control the motor coils. The solution has the advantage of a small dissipated power both in continuously and in switching regime.

The transient time for the current rising through the motor coils is proportional to the supply voltage. Initially, the motor is supplied with a voltage of 24 V higher than the standard rated voltage of 12 V. When the circuits reach the rated current, U1 and U6 decrease the pulse filling factor applied on the transistor grids by means of specialized circuits U5, which produces the stabilization of the current through coils.

This solution allows to increase the frequency of steps and to reduce the probability of losing steps. At the same time, the power dissipated by the output transistors Q1–Q4 is reduced, because the transistors work in blocked mode - saturated.

Potentiometer R12 allows the current applied to the motor to be adjusted.

Figure 10 and Fig. 11 respectively, show the printed circuit and the positioning of electronic components on the STEP board.

Figure 12 shows the connection of the STEP module to the acquisition board and to the stepper motor.

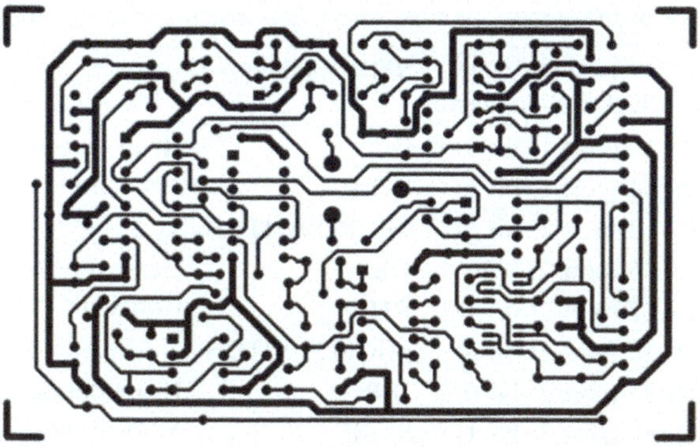

Fig. 10. The printed circuit of the STEP board.

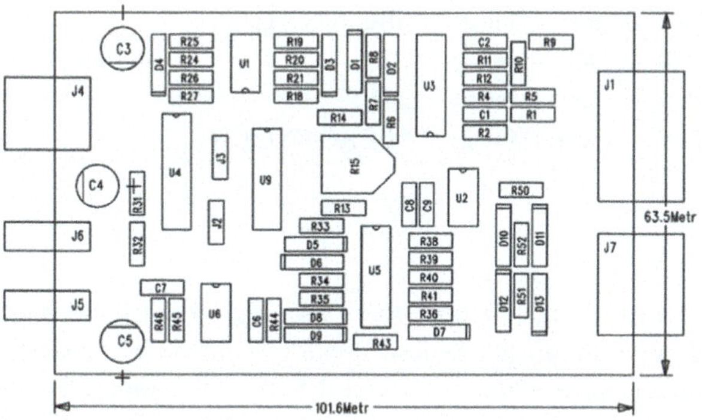

Fig. 11. Positioning of electronic components on the STEP board.

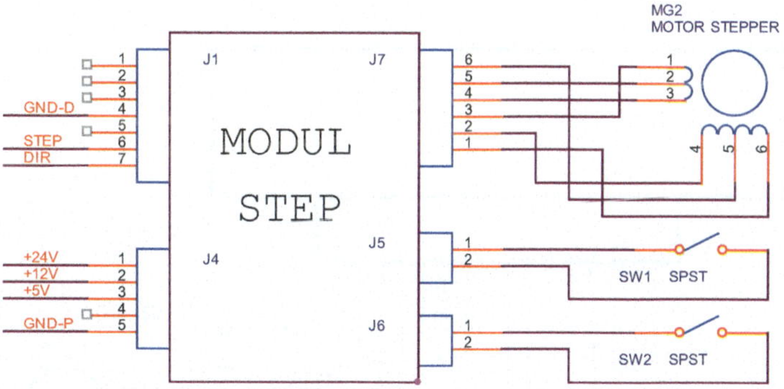

Fig. 12. Connection mode of the STEP module.

To control the five stepper motors, the acquisition board is connected according to the table below (Table 1):

Table 1. Control of the stepper motors, in correspondence with the ordered movement

Data acquisition board DIGITAL I/O connector	STEP module	
PIN 10 (port 1: bit 0)	DIR	**Head rotation**
PIN 44 (port 1: bit 1)	STEP	Vertical plane
PIN 12 (port 1: bit 3)	MICROSWITCH	
PIN 16 (port 2: bit 0)	DIR	**Head rotation**
PIN 17 (port 2: bit 1)	STEP	Horizontal plane
PIN 52 (port 2: bit 3)	MICROSWITCH	
PIN 23 (port 3: bit 0)	DIR	**Eye rotation**
PIN 57 (port 3: bit 1)	STEP	Vertical plane
PIN 25 (port 3: bit 3)	MICROSWITCH	
PIN 29 (port 4: bit 0)	DIR	**Eye rotation**
PIN 63 (port 4: bit 1)	STEP	Horizontal plane
PIN 31 (port 4: bit 3)	MICROSWITCH	
PIN 32 (port 4: bit 4)	DIR	**Eyelid rotation**
PIN 66 (port 4: bit 5)	STEP	
PIN 34 (port 4: bit 7)	MICROSWITCH	

SERVO Module

The control signals of the acquisition board are amplified by the specialized modules LSC 30/2 (linear servo controller in four quadrants) made by Maxon. The connection of this module to the data acquisition board and to the motor is shown in Fig. 13. The GND, CHA and CHB signals from the output of the servo motor encoder are also used by the acquisition board.

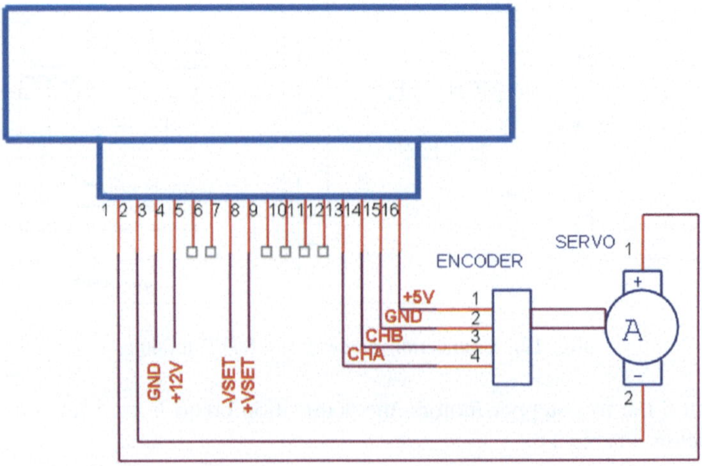

Fig. 13. Connection mode of the SERVO module.

4 Conclusions

The designed robotic head has eight degrees of freedom. They allow the simulation of specific human actions, in order to transmit information and to express positive emotional states, during the interaction with a human subject. Actions arising from these requirements are performed with mechanisms and elements of actuation and control, which are programmed in accordance with the voice message to be played.

The degree of similarity of the robotic head with the human facies depends mainly on the reaction speed of the robot, influenced through the quality of the used motors and the electronic control scheme, determining elements for the robot's simulation capacity.

References

1. Aryan, an Interactive Robot Face. https://people.csail.mit.edu/hmobahi/aryan/. Accessed 10 May 2020
2. What is Robonaut? https://www.nasa.gov/audience/forstudents/k-4/stories/nasa-knows/what-is-robonaut-k4.html. Accessed 15 May 2020
3. Barciela, G., Paz, E., Lopez, J., Sanz, R., Perez, D.: Building a robot head: design and control issues. In: Proceedings of the 9th "Workshop en Agentes Fisicos" (WAF 2008), Vigo, Spain, pp. 33–39 (2008)
4. Asheber, W.T., Lin, C.-Y., Yen, S.H.: Humanoid head face mechanism with expandable facial expressions. Int. J. Adv. Rob. Syst. **13**, 29 (2016). https://doi.org/10.5772/62181
5. Asada, H.: Introduction to robotics, 2.12 Lecture notes, Department of Mechanical Engineering, Massachusetts Institute of Technology. http://people.csail.mit.edu/jbarry/spring2011PR2/readings/asado.pdf. Accessed 01 Mar 2020
6. Udrea, C., Alexandrescu, N., Panaitopol, H., Avram, M., Apostolescu, T.C.: The Constructive Bases of Industrial Robots. Editura Universitara Publishing House, Bucharest (2006). (in Romanian)
7. Apostolescu, T.C., Ionascu, G., Petrache, S., Bogatu, L., Cartal, L.A.: Electromechanical structure of the experimental model of a robotic head. In: Gheorghe, G. (eds.) Proceedings of the International Conference of Mechatronics and Cyber-MixMechatronics, ICOMECYME 2019. LNNS, vol. 85. Springer, Cham (2020)

Educational System for Augmented Reality Applications

Cristian-Gabriel Alionte, Alexandru-Hanni Al Shehari, and Liviu-Marian Ungureanu(✉)

University Politehnica of Bucharest, Splaiul Independentei 313, Bucharest, Romania
liviu.ungureanu@upb.ro

Abstract. In this article, we present a mechatronic education system need it by students to learn the application of the augmented reality. Using a tablet, the user can identify any equipment and can control and modify the working parameters. Using the software which use augmented reality on our demo mechatronic education system, the tablet identifies the motor, the switches, and the converters. Also, on the tablet screen are displayed the working parameters and some of them can be modified by user using a simple touch on the parameter.

Keywords: Mechatronics · Augmented reality · Neural networks

1 Introduction

1.1 Industry 4.0

Given that Industry 4.0 is still at a conceptual stage and that its implementation requires to include a complex and technologically dynamic process that will include multiple industries such as IT, electricity suppliers, construction, medicine etc., will make it difficult to expand information globally, a process that is absolutely necessary in making the transition from third industrialization to the new type of industrialization.

Digitization of industry is a step that is still in the working, but technologies such as artificial intelligence, big data and new connectivity methods indicate the certainty of a new digital revolution. Industry 4.0 is what will have a strong impact on the complete transformation of the industry precisely because of the three key points of progress they represent:

- Productivity digitization – new information management and production planning systems
- Automation – efficient data acquisition systems directly from production lines, which is possible due to machine retrofitting
- Automatic exchange of data – linking manufacturing sites to supply chains by means of rapid exchange of information between the two.

The basic feature of this new method of industrialization lies in increasing competitiveness in the "intelligent or smart" equipment market but also the use of information which are not essential from the user point of view but are import for sustanable

G. I. Gheorghe (Ed.): ICOMECYME 2020, LNNS 143, pp. 72–80, 2020.
https://doi.org/10.1007/978-3-030-53973-3_8

development (demographic changes, resources, energy efficiency and the level of urbanization or production of an industry).

The four key concepts of this industry are cyber-physical systems, systems such as Virtual Reality, Augmented Reality or even holography that are methods of easily connecting virtual reality with everyday reality, the Internet of Things, the Internet of Services and smart factories. Another important point to note is that communication between machines and "intelligent or smart" equipment will never be an independent object of study but, on the contrary, one dependent on another concept of resonance for what we call Industry 4.0.

1.2 Augmented Reality

Augmented Reality can be defined as a direct or indirect view of the environment that has been improved by adding virtual applications generated by a computer. AR is an interactive environment, providing a real, 3D experience – thus combining real and virtual objects.

The augmented reality aims to simplify the user's life by exposing information through virtual means, not only for its immediate surroundings, but also for any indirect view of the real-world environment, such as live-video streaming. This concept improves the user's precept and interaction with the real world. The technology also offers enormous potential, which can also be applied to human senses, thus increasing smell, touch, and hearing.

It can also be used to increase or substitute missing user senses through sensory substitution, such as increasing the visibility of blind users or users with poor visibility using audio indices. Information transmitted by virtual objects can help the user in carrying out daily tasks, such as guiding workers by electric wires in an aircraft by displaying digital information through a headset. For this kind of technology there is an abundance of applications such as medical visualization, entertainment, advertising, planning a route for robots, etc.

Augmented reality faces many technical challenges, such as limitations on the use of a stereo viewing device with high resolution, varied color palette, power brightness, contrast, and focus power.

One of the common limitations of this technology is portability and its use in an uncontrolled space. This limitation is exposed due to the connections between all devices that must be able to withstand a combination of external risk agents such as high or very low temperatures, high humidity, etc. By contrast, with the help of new IoT technologies, all data is easily transported from a safe place and the only product that must withstand all these problems is the augmented reality mechanism (mobile phone, tablet, laptop or smart glasses).

Another limitation to be mentioned is image quality and rapid adaptation capacity. From this point of view, the technology needs to be developed to provide extra quality for all the images it overlays. So, when a video camera must use an augmented reality software, it submitsun double effort for processing 3D objects to be superimposed but also on the realities that must or express with a much higher quality.

One of the most problematic limitations is that of social acceptance of this technology. Being a technology that at first was quite inaccessible, being used heavy and expensive equipment, at the social level a rather complex conception of this technology was formed. Now that technology has advanced it is much easier to be integrated into different applications as the mobile phone is one of the most used IT structures of the moment.

2 Mechatronic Education System

In this article we present an education system need it to demonstrate the augmented reality.

As far as the practical part is concerned, a mobile system made up of industrial equipment will be provided with the aim of exposing all three areas that mechatronics offers from the beginning of its definition, namely: the mechanical, electronic and automation parts.

The mechanical part is represented by two frequency converters, connected in parallel with an alternating current motor. The purpose of the mechanical application is to demonstrate the different actuating methods of the two frequency converters and their way of controlling and controlling the rotation speed of the synchronous engine, adjusting the frequency and size of the engine power voltage. The two converters are exposed precisely for the different applications in which they can be used, one of which has specialized crane application control settings and the other has specially designed options for pump running.

The electronic part consists of a PLC – Programmable Logic Controller, which will aim to purchase data from both converters simultaneously, using the Modbus protocol. After data acquisition, it will be processed and brought in a simpler form to use and read to reduce data transmission times in the cloud. The latter operation will be carried out using the visual programming environment.

The automation part includes both programming the PLC and the data collection through another programming medium called Node-Red, making a local server and then transmitting the data through an Application Programming Interface, which will identify and modify the data in real time on a tablet running our own Augmented Reality software.

As represented in the logic scheme of the previous system, the educational prototype consists of a PLC, two converters, an engine, and a router. The converters on display are part of two different ranges, each of which has different advantages and disadvantages of cost operation, and energy saving.

The router will aim to create a common network between the data storage and the educational prototype so that data can be exported from the data storage. The user computer or tablet or smartphone will have multiple roles such as:

- Building the augmented reality application and running it through a controlled programming environment – Python and JavaScript
- Processing augmented reality simulator elements using a dedicated software for direct reading and transmitting information through an API – "Application programming interface"
- Retrieve data from co-vertigos and convert it to a preset MQTT format

- Transmitting data to the cloud server and ensuring communication between the tablet that will run the augmented reality app and the cloud (Fig. 1).

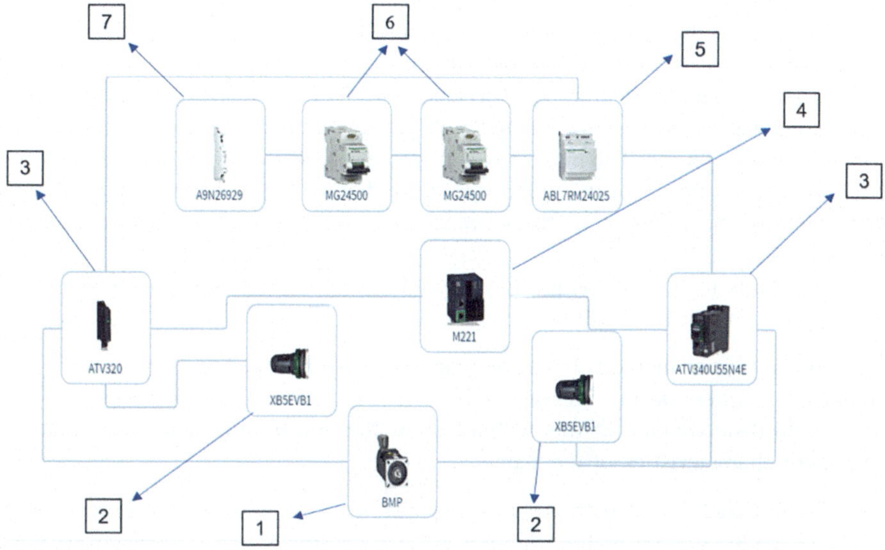

Fig. 1. Architecture of the mechatronic system: 1. Synchronous engine, 2. Potentiometer, 3. Frequency converter, 4. PLC, 5. Source, 6. Switch, 7. Auxiliary Contact.

In the realization of the augmented reality software, they were used from complex neural systems to related programming languages that led to the understanding, use and efficiency of open-source software for reprogramming - open-source detection of objects in space without markers and that can help complete the augmented reality software, with the ultimate aim of detecting objects using a camera and overlapping virtual objects on this image.

In augmented reality, applying knowledge of position and orientation from the perspective of the camera (or camera position) is essential because it ensures those of the raised or augmented stage. Indeed, the position estimation allows the consistency of space-time to model a virtual camera through which to make a rendering of the virtual world with the same features of its camera. Proper alignment of real and virtual worlds. There are two main classes of photo estimation methods: method-based markers and method less markers.

Method-based markers place artificial targets in the real scene to facilitate visual tracking and estimation. Bullets are detected in a simple way, containing codes that they distinguish. Their positions in the reference world are unknown a priori and the estimation of their position follows the following scheme: detection of markers, 2D/3D mapping and, finally, calculation of camera position.

Methods without markers exploit the natural characteristics existing in the real scene in the form of corners edges and line segments. Data extracted from the 2D image of the scene is mapped to 3D data extracted from the 3D model of the scene.

To increase the user's real environment, we need to have a camera mounted in relation to a landmark. Its calibration is to determine, geometrically, its optical properties, its position and orientation. In other words, it is to calculate the picture (3 orientation parameters and 3 position parameters) of the actual camera to make a "coincidence" with the virtual camera (the one used for 3D perpetual playback). To do this, the most used method for indoor AR applications (prepared environment) is to place real scene markers that are used to calculate 3D coordinates from three specific points recognized by the system. The technology technique can be used with a webcam or a simple mobile phone. The marker-based growth process consists in the sequence of operations on each image of the video stream to identify the presence of a marker and then to identify the different markers uploaded to the application (in the context of multiple markers).

Although this technology is the most widely used and best known for augmented reality applications, it has limitations:

- A disadvantage of methods based on markers is to be limited to their area of vision. Indeed, a marker can become detectable when the camera is removed.
- The main disadvantage of this technology is the non-homogeneity and change of the main characteristics of the scene.

Unlike approach-based markers, in this paper we want to allow the user to test an augmented reality system without having to have capture systems or electromagnetic motion markers.

Therefore, the main purpose is to achieve an augmented reality system based on the camera of the computer/tablet/smartphone. We are particularly interested in increasing the component on the real stage of this scene.

In general, using machine learning software require large data sets to obtain a good model. The biggest problem, besides the bandwidth saturation and big data storage necessity, consists that it can be complicated to get such a large data set because, usually, labels with that data need to be entered manually. These tags will be used by the neural network to verify that the forecast values are correct or not.

Also, instead of allowing the software to use only original color images, many edited images were used for better results by rotating, twisting, changing their brightness, saturation, contrast, etc. This will essentially be an image of the same object, but the data will look different, thus giving the data model more relevant, precisely to prepare different recognition scenarios.

Designing a neural network for image classification was not the easiest task and a lot of effort is to try and replace many pieces of code until the success rate comes close to the best use experience. For our educational system, the scope was achieving a success rate of over 90%. Thus, for the proper definition of the project we started by running several convolutive layers, the final layer having four nodes and several Softmax activation function [6] to give a probability of classification between four classes. The activation functions of the other layers have been set to Relu function [6] because they are much faster and preferable for higher-sized networks.

The Relu function (the corrected linear unit function) is a good option when negative values should be ignored or make no sense for that code. It is also a very good choice when it is intended to avoid process differences and increase the level of processing due to the increase in the level of calculation, this being done because the function offers similar values between input and output but also sets all negative values to zero. This function is usually a good choice in the construction of larger neural networks find used.

The most important aspect to mention is the difference between the number of photos processed and how they were taken. Adding more photos made a huge difference to the test rating score. After repeatedly adding more images, the test score gradually increased from 80.55% to 83.45% in accuracy. The model was already trained on the original image and the result was sufficiently similar, so that because of certain errors there were some unrealistic results of 99.52%. However, this was remedied by dividing test data and other data into separate folders (Figs. 2, 3 and 4).

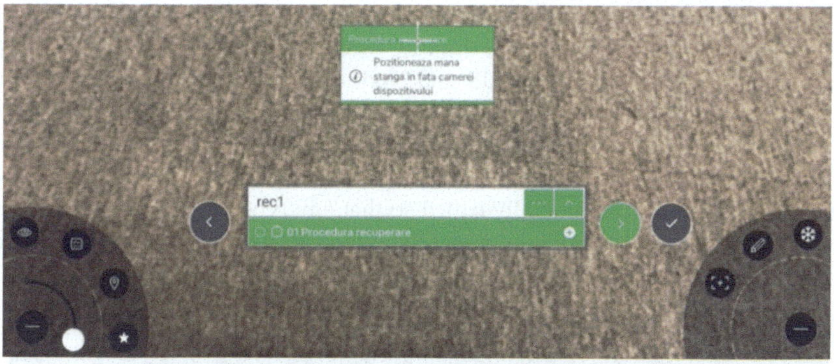

Fig. 2. Photo of the completed and fully functional app with instructions for use

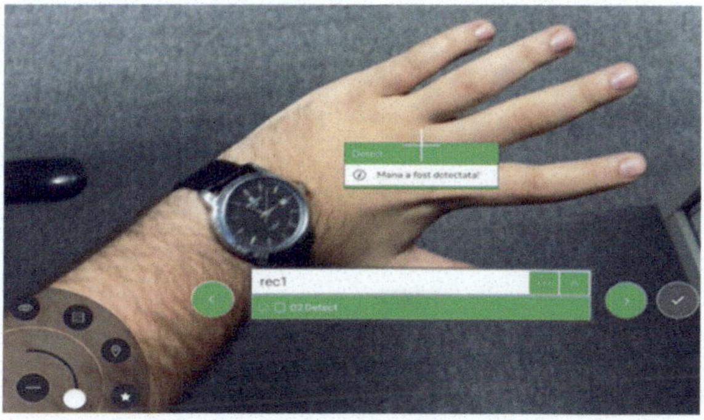

Fig. 3. Picture of the completed and fully functional app – hand detection

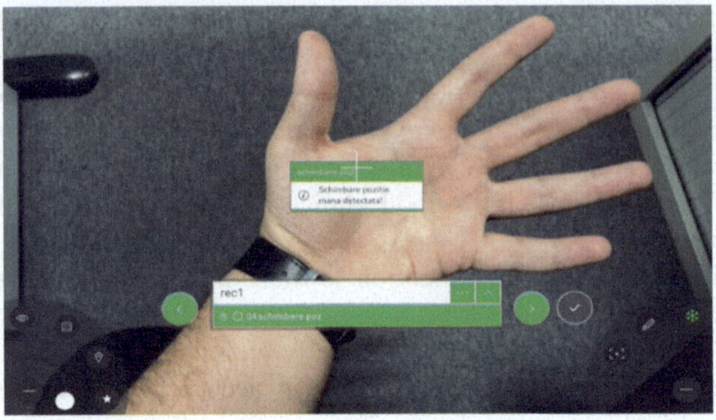

Fig. 4. Photo of the completed and fully functional app – detection of hand position change

The connection between the equipment and the augmented reality platform, which will aim to detect industrial equipment, will be made using an API – Application Programming Interface made as an object in the Node-Red platform using JavaScript.

In terms of hardware connection, the PLC was connected to a router via ethernet. Further a laptop on which the NodeJS and Node-Red software were installed. It was necessary to develop two virtual programming objects, the Node-Red platform using these virtual objects connectable to each other to facilitate the programming method for a complete IoT system. So the platform at the time of installation offers programmable or node objects, as they are officially called so that any type of user can build a desired IoT architecture for any type of equipment, be it industrial or addressed to smart homes (Fig. 5).

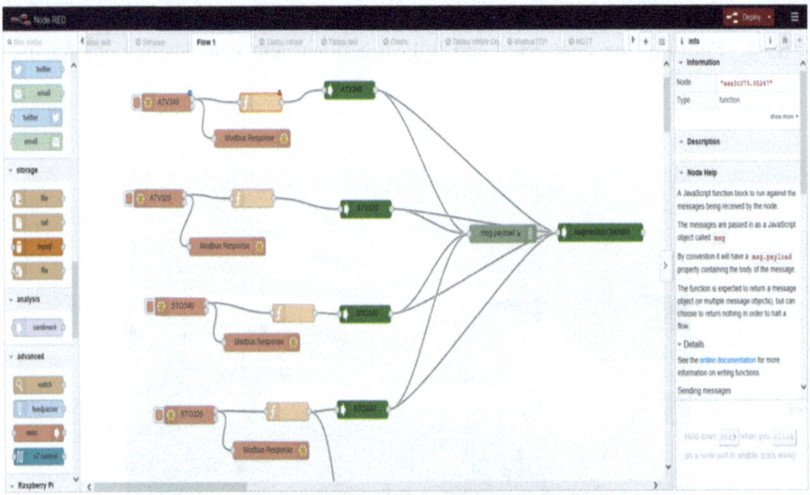

Fig. 5. Node-Red platform – final data transmission architecture

These objects can be programmed to retrieve information through different types of protocols, in the case of this project it will be only the Modbus protocol, and for the connection between the PLC and the computer will be entered the IP of the PLC from which the reading is performed, the reading port and the reading address already defined within the developed software will then be able to read the data from the PLC at a time interval defined by the user as follows. as shown in the following image.

After the actual programming of the objects responsible for taking the information from the PLC for each defined reading address, one engine operating frequency and one STO parameter from each frequency converter, the data read by the objects was forwarded to the two special information formatting objects for the augmented reality application performed.

The first special object is the one where the object specifies the route where the information should be displayed within the augmented reality application. Specifically, to display information about the engine's running frequency, we specified a random name in the augmented reality software, specifying the same name in the edited object in Node-Red so that it knows that the data read must be displayed on the default route.

The second special object, is the discussed API itself, which provides the ability to connect to the augmented reality application and display the information that the platform reads in PLC.

3 Results and Conclusions

Finally, we wanted to present several explanatory images of the final solution resulting from the application of all the mathematical methods described above, this paper having as main topic the presentation of the methods of implementation and use of industry 4.0 technologies in a mechatronic system. This was done with the help of connected industrial equipment for transmitting data in a virtual environment. All this industrial equipment has been incorporated into a suitcase to ensure the mobility of the solution presented in this paper (Fig. 6).

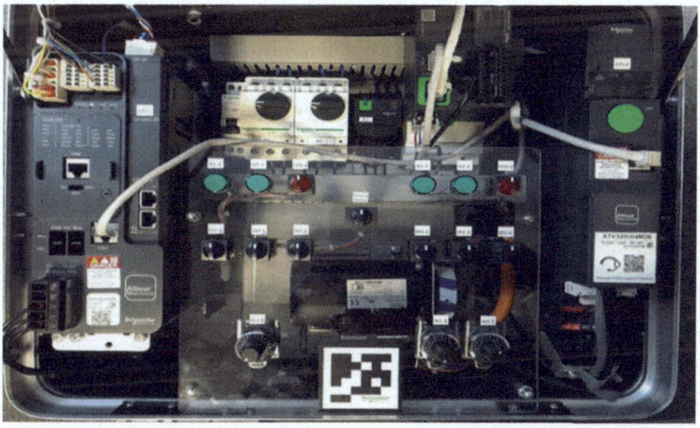

Fig. 6. Presentation of own solution – demonstration equipment

After the application is fully installed and images are configured to be enhanced with relevant data such as circuit scheme, presented equipment user manuals, or real-time data obtained directly from the PLC (Fig. 7).

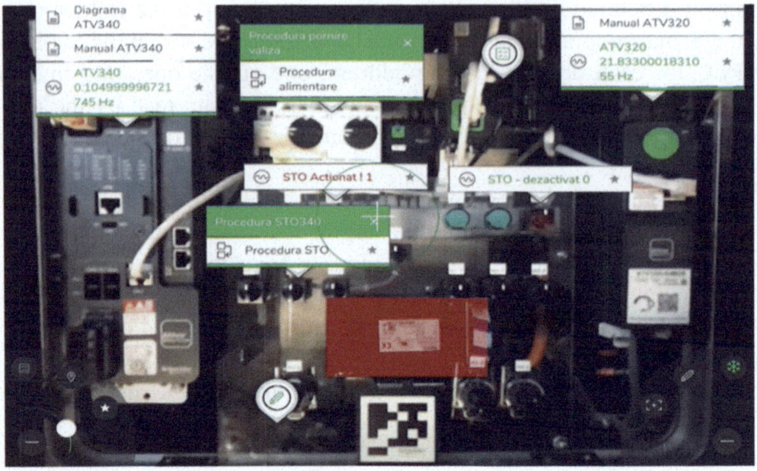

Fig. 7. The educational prototype viewed in functional application which run on a Samsung Galaxy Tab S6. It can be seen the presentation of parameters in the augmented reality module on the tablet

References

1. van Krevelen, R.: Augmented Reality: Technologies, Applications, and Limitations. VU University, Amsterdam (2007)
2. van Krevelen, D.W.F., Poelman, R.: A survey of augmented reality technologies, applications and limitations. Int. J. Virtual Reality **9**(2), 1–20 (2010)
3. Billinghurst, M., Kato, H., Poupyrev, I.: The MagicBook-moving seamlessly between reality and virtuality. Comput. Graph. Appl. **21**, 3 (2001)
4. Buchmann, V., Violich, S., Billinghurst, M., Cockburn, A.: FingARtips: gesture based direct manipulation in augmented reality. GRAPHITE 2004: Proceedings of 2nd International Conference on Computer Graphics and Interactive Techniques. ACM Press (2004)
5. Augmented reality gets real. Commun. ACM (2019). https://www.doi.org/10.1145/3344293,62,9,(16-18)
6. Chollet, F.: Deep Learning with Python. Manning Publications Co. (2018)

Snow Mobile Robot - SnowBie

Cristian Gabriel Alionte, Ciobanu Alexandru Costin, and Liviu-Marian Ungureanu(✉)

University Politehnica of Bucharest, Splaiul Independentei, 313, Bucharest, Romania
liviu.ungureanu@upb.ro

Abstract. In this paper we present the SnowBie, the snow mobile robot. We start the design and the construction of the robot because we want to show a new type of non-skid wheel in an easy and chcaper way.

Keywords: Mobile robot · Mechatronics

1 Introduction

The project has as the scope of presentation of a robot capable of moving in rough terrain with the help of the adaptive wheel system. Through rough terrain, in this project, we understand a snow-covered terrain that are the properties: it is skidpery, and the snow prevents the identification of the distance between the robot and solid ground.

This caused the robot to have a chassis as light as possible but with as much surface area as possible so that the snow cover could support the robot's weight. The movement can be done by implementing a wheel-level system that allows it not to skid.

This robot brings a benefit to the research area that requires such systems capable of coping with environmental conditions. The robot will be able to move on different types of terrain such as: sandy terrain, snowy terrain, muddy terrain, pole-covered terrain, etc. It will be able to bypass obstacles and will be remotely controllable via a Bluetooth module.

The mechatronic system presented in this educational project is only at the prototype stage to determine its reliability, robustness, and effectiveness. The new adaptive wheel system can be independently taken over and used to modify the current wheel-shifting system for any wheel drive equipment starting from robots and up to vehicles and beyond. This change would bring several advantages, namely:

- driver-controlled anti-skid system;
- exclusion of the time expensive to installation/dismantling of the anti-skid system;
- maintaining the driver's comfort while driving on the road;
- excluding the need for technical knowledge and the difficulties that may arise in the system;
- maintaining a small system gauge (only the outer wheel valve is modified);
- the ability to move on several types of terrain and especially on snow or ice.

Several anti-skid solutions are known at the moment, of which we recall.

G. I. Gheorghe (Ed.): ICOMECYME 2020, LNNS 143, pp. 81–89, 2020.
https://doi.org/10.1007/978-3-030-53973-3_9

Unconventional tire chains [2] shown in Fig. 1a. The main advantage of the chain-skid system is the low cost compared to other systems of this type. The simplicity and efficiency of the system is the reason why these chains are so widespread. The major disadvantage is the installation difficulties. Chains cannot always be mounted on the vehicle, especially when it is travelling on the road. Skidding usually occurs at unexpected times, which involves mounting in difficult positions, sometimes impossible for certain people. Another major disadvantage of this system is the time wasted for mounting/dismantling the chains.

Chain mechanism for a vehicle on snow [3] shown in Fig. 1b and c. Another example of an anti-skid system refers to an automatic mechanism with radially arranged chains. During snow, the classic snow chain system is installed on wheels to prevent them from skidping and for traction. Typically, snow chains are mounted around the outer tread of the tire. Friction between the chains and the road surface prevents the wheels from skidping. The anti-skid mechanism for motor vehicles developed [3] comprises a set of radially arranged chains on a disc connected to an electric motor. The electric motor is attached to an arm that helps to rotate the assembly completely if it is actuated. The equipment shall have a control module so that the positioning of the chains between the wheels and the roadbed is carried out when the driver is required to operate the mechanism. This is very beneficial when the vehicle is moving on different surfaces such as snow, asphalt, muddy terrain etc. In the Fig. 1b is presented such a system that counteracts the skidding, mounted on the 2th chassis of a motor vehicle. The chain mechanism includes an engine 6 for rotating a metal disc 8 on which a multitude of chains 12 are mounted. In a preferred variant, engine 6 is supported by a second articulated arm 18 mounted under the 2. Part of engine 6 is coupled to cylinder 16. Cylinder 16 is the one that causes the subassembly to rotate with chains, depending on the command given by the driver. In the Fig. 1c is presented the rotary plate 8 to which are coupled a multitude of chains 12 arranged radially. During use, when rotary plate 8 is driven by engine 6, chains 12 are broken down due to the centrifugal force that appears and causes them to spread under the wheels, thus preventing the tires from sliding on snow or frozen roads [3].

Anti-skid apparatus for disc vehicles [4]. A schematic view of the rear portion of a car, with the anti-skid elements raised, in the resting position (Fig. 1d) and in the operating position (Fig. 1e). The advantage of this method is the system control mode, which is switched off and operated at the driver's preferences. [4] Mechanical arms 4, which switch between actuated/resting positions, have rigidly caught at the ends of discs 3, which rotate under the traction forces of the tires. The discs are made of rubber or similar flexible materials, reinforced with ropes or fabrics and are slightly curved near the periphery. Each disc has a central hole for the screw that rotates the disc. Disk-carrying arms are manually operated by the driver and switch between the positions in the Fig. 1d and e. When the disc is operated to be coupled to the edge of the tire tread, its edges, beveled and thickened, will be pressed tightly onto the roadway by the hemmed edge of the tire tread. The disk will be automatically locked in that position due to the 3.5-lane dreaded running area. [4] The major disadvantage of this invention is the size of the system. The vehicle on which the system is mounted must have a ground clearance large enough to allow the installation and operation of the system.

Another disadvantage is the rubber disc channels which can be clogged with various materials and lead to poor operation of the equipment.

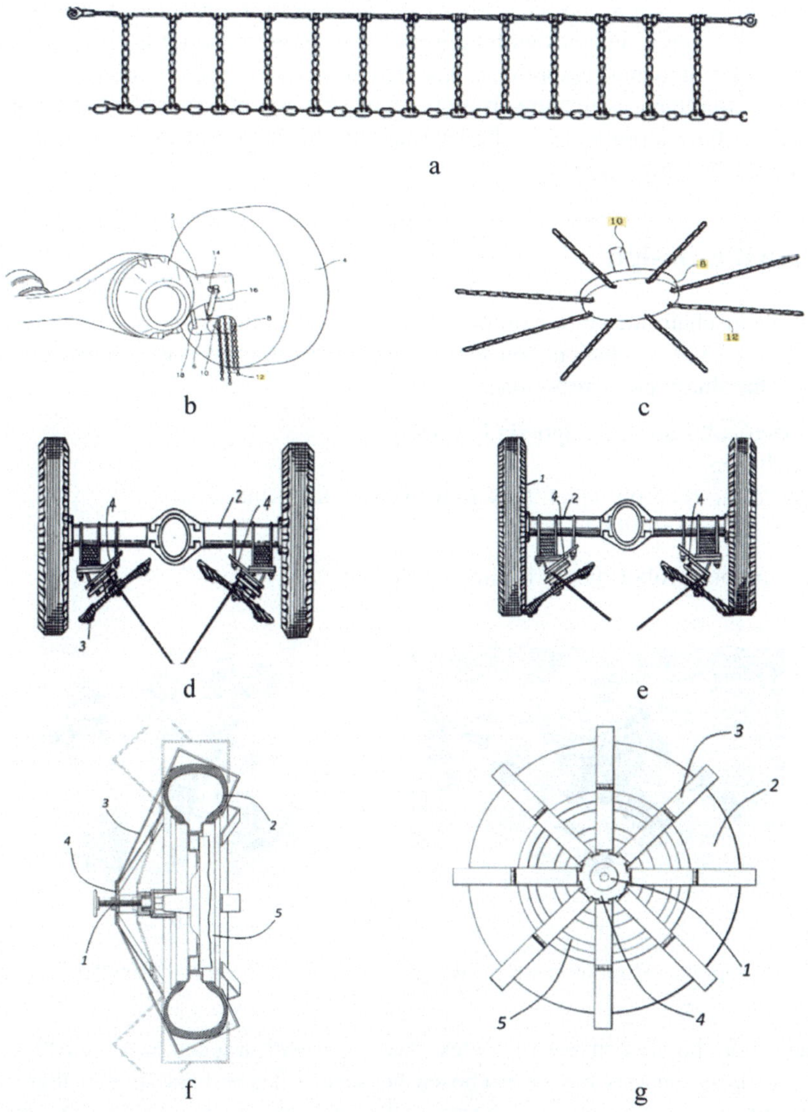

Fig. 1. Anti-skid systems

Anti-skid device with retractable arms [5] in Fig. 1f and g. The main purpose of the invention is to provide a device that can be easily applied to car tires without the need to lift the wheel. The articulated arms 3 are radially arranged on the wheel of the vehicle being articulated with the grip disc 4. The latter has a grip and screw-nut advance 1. An important feature of the equipment is the “U” shape at the end of the articulated arms 3. This allows effective grip on the 5th, i.e. a large tightening force and providing a sharp anti-skid surface that couples on the road. [5] It should be noted that when put into operation, the adhesion of the vehicle to the ground increases considerably but, if it moves in areas with high levels of snow or sand it is no longer able to cope with environmental conditions. Another disadvantage is the joints between the arms of the equipment which are strongly required externally (bending, traction, etc.). This leads to the deformation of the bearing and the failure of the system (it can no longer be folded properly) [5].

2 SnowBie Design

The mobile mechatronic system proposed in this project aims to move a robot on a field with a low friction coefficient (on ice) or on difficult terrain (sand, snow, etc.). It is intended that this type of robot has:

- an electrically operated anti-skid system
- a small size
- adapt to the road on which they move (snow, ice, sand, etc.)
- be remotely controllable.

The proposed robot model is shown schematically in Fig. 2.

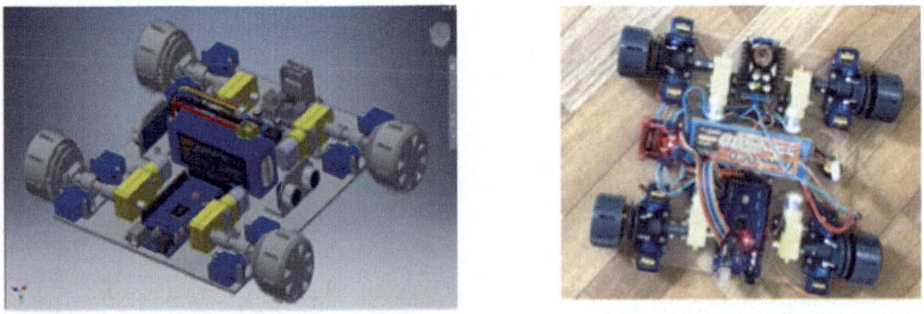

a b

Fig. 2. 3D robot model. a) design of the robot; b) the real prototype

The robot chassis consists of a plexiglass plate with a gauge size of 220 * 240 * 3 mm. A plastic chassis has been chosen because it has a mass smaller than that of metals and is easier to process mechanically. The processing technology that was chosen was laser cutting.

The main part of the structure is the anti-skid wheel. This contains a complex adaptive mechanism which allows the antifriction teeth to exit the upper margin of the wheel.

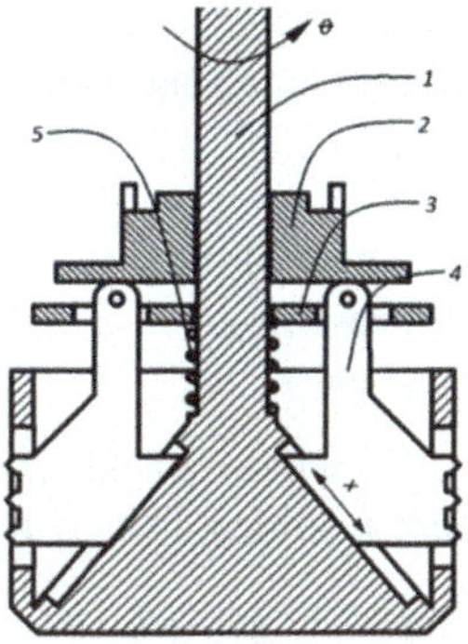

Fig. 3. Anti-skid wheel

Corps 1 is intended to come into contact with the surface on which the robot is moving (road, sand, ice) and is driven to one end by a DC Motor as shown in Fig. 3. Discs 2 and 3 are placed on the tree of element 1 and are intended to train the toothed pallets 4. Disk 3 has a slot for each 4 toothed palette and rotates with body 1 when the DC motor is powered. When the RC Servo mechanism (Fig. 3) receives a command signal, it pushes disk 2, which in turn pushes the 4 pallets. The latter translates to the sloping plane and alters the outer surface of the wheel. Disk 2 is fixed, having two sliding channels for the arms of the servomechanisms. This causes the heads of palettes 4 to be freely rubbed on disc 2. The helical arch 5 is intended to bring and maintain the blades in the initial position when the arms of the servomechanisms are withdrawn.

The coupling between the shaft at the exit of the engine reducer and the wheel shaft is shown in Fig. 3. The coupling is a rigid one, made by rapid prototyping technology. Two screws shall be used as a fastener.

The following calculation scheme was drawn for the choice of the type of servo mechanism. It is a wheel with toothed pallets. A single pallet has been taken in contact with the ground.

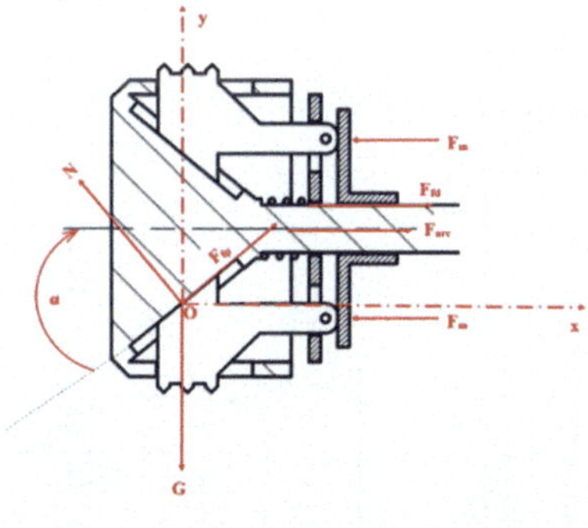

Fig. 4. Calculation scheme of the anti-skid wheel

To simplify the calculation, choose the convenient XOY coordinate system, as shown in Fig. 4. The following forces appear in the system:

F_m engine drive force
F_{fp} friction force at skid between blades and wheel
G robot weight on one wheel
Normal n on the surface
$\alpha = 45^o$ angle of inclination of the sliding surface
F_{arc} compression force that appears in arc
F_{fd} friction force between discs and shaft

To determine the action force F_m will be made a steady study of static forces. The arc is compressed to the maximum resulting from a force:

$$F_{arc} = F_{a0} + k * f \tag{1}$$

where:

- F_{a0} is the spring pretension force
- k is the stiffness of the arc
- $f = 5$ mm is the maximum race that the tooth palette makes

The pretension force of the arc is given by the relationship:

$$F_{a0} = \frac{f_0}{f - f_0} * F_{arc} = 0.222\ N \tag{2}$$

The rigidity of the arc is given by the relationship:

$$k = \frac{F_{arc} - F_{a0}}{f} = 1,9556\ \frac{\text{N}}{\text{mm}} \tag{3}$$

The length of the semi-manufacturer for the helical spring shall be calculated with the relationship:

$$l = \frac{\pi * n_t * D_m}{\cos \alpha_0} = 483.46\,\text{mm} \tag{4}$$

where:
$\alpha_0 = 6°$ is the angle of tilt of the free-range

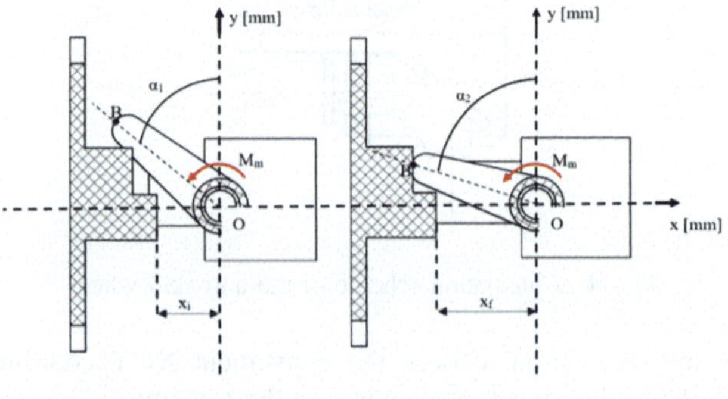

Fig. 5. Angular displacement of the servo arm of the mechanism

It was chosen to operate with two servo mechanisms to reduce the friction force between the channel disc and the shaft. On the two channels of the disc are leaning arms of the servo mechanisms (Fig. 3). To choose the optimal model of servo mechanism should be known the following:

- Force F_m
- Rotation angle

The force Fm servomechanism is given by the relationship (6) of the system of:

$$F_m = \frac{F_{arc} + F_{fp} + F_{fd} - N * cos\alpha}{2} = 2.0235\ N \tag{5}$$

The angle that the servo arm makes to the mechanism with the disc in the initial position is $\alpha_{1.}$ and the disc is in position x_i. When the servo mechanism is operated, the arm rotates in a trigonometric direction until the disc reaches the final position x_f, i.e. after the disc has moved by 5 mm:

$$\alpha = \alpha_1 - \alpha_2 = \arcsin\frac{x_f - x_i}{OB} = 19.4721^\circ \tag{6}$$

where:

- *OB* = *15* mm is the length of the arm of the servomechanism

The motor moment of the servomechanism is given by the:

$$M_m = F_m * OB = 2.0235 * 15 = 30.3525\ \text{Nmm} \tag{7}$$

The working angle has a maximum value of 270 or 3o/2 rad.

The RC servo has a stroke of about 180° or α rad and therefore this range will be centered on the middle of the potentiometer (3o/4). The command for the middle position (0° angle to the engine shaft) will be given by a signal with a duration of 1 logical of 1,5 ms and a break of 20 ms. For pulse durations of less than 1.5 ms, the servo rotates clockwise (inversely trigonometric), and at 0.5 ms reaches the maximum position for this purpose (90°), detected with the reaction of the potentiometer (Fig. 5).

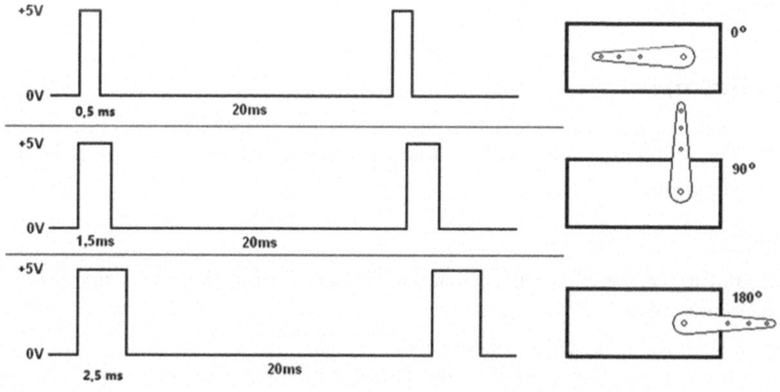

Fig. 6. Command signal for servo mechanism

The specified 20 ms interval is the maximum range of transmission of the control pulses of a mechanism servo so that the movement is continuous. For pulse durations greater than 1.5 ms, the servo rotates counterclockwise (trigonometrically), and at 2.5 ms reaches the maximum position in this direction (−90°), a position detected with the reaction of the potentiometer (Fig. 6).

The parts made by the rapid prototyping process are shown in Figs. 7a to d. This type of processing has been chosen because the chosen constructive solution is of a prototype nature. However, the pieces have a durable structure and a complex design. This processing technology is easy to use and capable of making structures with heavy or impossible geometries on classical processing machines (strung, mill, etc.).

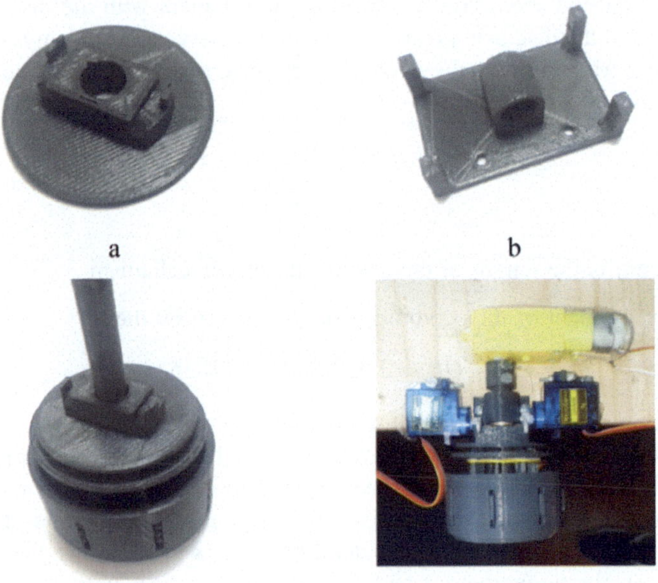

a b c d

Fig. 7. Anti-skid wheel: a) disc with channels; b) radial camp support and servomechanisms; c) adaptive wheel subassembly; d) test support

3 Conclusions

The following advantages and disadvantages have been identified as a result of the practical realization:

Advantages:

- Accumulating practical work experience by performing the necessary processing to assemble the robot;
- Create a system with real applications;
- Flexibility and creativity in solving various problems arising during design and assembly;

Disadvantages:

- The emergence of more or less high costs of purchasing parts, material for the 3D printer and processing parts;
- Breaking of brittle components (wheel shaft and camp holder).

Practical achievement has been a set of necessary experiences combined with challenges that have arisen throughout the design, from the implementation of the idea designed schematically in a 3D model, to the choice of materials and parts that are part of the system.

References

1. Choi, J.-S.: Chonrabuk-do (KR), Snow chain mechanism for a vehicle
2. Mank, R.A.: Pittsburgh, Pa. Application, Serial No. 340,046 1. Claim. (Cl. 152–213) The chains, 3 March 1953
3. Robinson, L.: Pittsburgh, Pa. Application, Serial No. 544,448 8 Claims. (CI. 188–4) Anti-skid apparatus for vehicles, 2 November 1955
4. Eisenhauer Sr., H.J., Buffalo, N.Y.: Application, serial No. 45,222 9 Claims, Anti-skid device, 20 August 1948
5. Siewert, J.M.: South Dakota, Antiskid attachment for wheels, Patented, 9 November 1920
6. Neson, A., Hinnard, H.: Non-skid wheel. application filed may 19, 1919, 1,886, 149 Patented, 6 April 1920
7. Kromer, G.J.: 2,608,274 Antiskid attachment filed, 27 February 1950
8. Gardner, N.C.: 2,754,874 Antiskid devices filed, 3 October 1952

Conceptual Model and Proof of Concept for a Complex Mechatronic System Used in Neuromuscular Control Training

Cristian Radu Badea(✉), Paul-Nicolae Ancuţa, Sergiu Dumitru, Anghel Constantin, and Nicuşor Nicolae

The National Institute of Research and Development in Mechatronics and Measurement Techniques INCDMTM, Pantelimon Street, 6-8, 021631 Bucharest, Romania
adresacontact@gmail.com

Abstract. The paper presents the conceptual model and the Proof of Concept of a complex mechatronic system which can executes predefined sequential motion cycles, used to create biofeedback reaction in order to improve the human neuromuscular control, the lower limbs muscles endurance and the coxo-femoral joints mobility. This mechatronic system can be an extremely useful tool in physical therapy, sports, medicine and rehabilitation. The system comprises three parts: a portable data acquisition and wireless transmission electronic module that performs complex analysis of the numerical values acquired from tactile force sensors, a stationary automation panel containing a Main Control Unit (MCU) and digital signal interfaces and a complex mechanical cinematic assembly consisting of solenoid valves, pneumatic cylinders and various mechanical parts. The numerical values of the reaction force are acquired from force signal conditioners, filtered and wireless transmitted to a stationary automation panel. The MCU electronic module is programmed to run a specific software application that ensures the proper execution of the motion cycles in order to obtain the feedback control from the human subject. This is due to the fact that the numerical signals and events issued or received by hardware components of the automation panel (digital inputs/outputs, operator panel action, wireless received data) has to be taken into consideration. This complex mechatronic system is a working progress research project.

Keywords: Complex mechatronic system · Biofeedback · Neuro-muscular control · Wireless real-time data transmission

1 Introduction

1.1 A Brief Description of the Mechatronic Equipment

The paper presents the conceptual model and the Proof of Concept of a complex mechatronic system which can used to increase the human neuromuscular control and to improve the lower limbs muscles endurance and the coxo-femoral joints mobility.

The mechatronic system will have the role of keeping the human subject into a continuously state of imbalance, this action provoking him to continuously find his

G. I. Gheorghe (Ed.): ICOMECYME 2020, LNNS 143, pp. 90–99, 2020.
https://doi.org/10.1007/978-3-030-53973-3_10

balance, which, in time, leads to the development of his neuromuscular control and lower limbs muscles endurance (or stamina). The human subject will have one foot (the manipulated leg), fixed on the effector element (called the trainer-arm), of the mechanical subassembly, and he will use the supporting leg (the foot in contact with the ground) to move in the direction imposed by movement of the trainer-arm.

The movement parameters for each session (the duration and the direction of the trainer-arm), will be controlled through the application software programmed in the MCU of the mechatronic equipment. By changing the duration of each session, an increase in the volume of effort, made by the supporting leg's musculature (of the human subject) is obtained.

The mechatronic equipment is intended to be used in it's testing phase together with an external motion analysis system that should deliver reports about the user's level of balance, which is one of the method used for determining the neuromuscular control of the human subject. These reports will also be used to calibrate and verify the electronic sensorial module's accuracy and also, they can be used to track and evaluate the improvement of each individual person's neuromuscular control, lower limbs muscles endurance and coxo-femoral joints mobility. The motion analysis system will be not described in this paper.

1.2 Previous and Related Work

Paper [1] presents a study that provides methods and results regarding neurophysiologic and kinematics regulation of postural responses during unilateral balancing by applying random, unexpected perturbations, which requires appropriate neuromuscular control to regain equilibrium after surface translation, as it is needed during slip-like conditions.

Compared to the above-mentioned work, the structure of our mechatronic equipment together with the concept of motion cycle allows the implementation of complex motions, that represent combinations of horizontal and vertical movement.

Paper [2] presents balance analysis through stability zones. The analysis of the stochastic variation of the center of pressure is carried out by integrating the values in an ellipse and determining the areas of stability and instability of human subject. The obtained mathematical equations define four regions: high preference area, low preference area, undesirable area (where the human subject had to change his posture to maintain his balance) and instability zone (where the subject is forced to apply a stabilization technique).

Paper [3] presents some results obtained by the correlated analysis (linear acceleration-trajectory recording) of the behaviour of the locomotors system of human subjects. Specialized equipment was used for the real-time acquisition of the linear acceleration values specific to the foot and leg. A virtual model for analysing the dynamics of the stability of the human subject that performs different actions.

In the above-mentioned works, evaluation and development of neuromuscular control is performed mainly with the human subject having a bilateral support posture (with both legs placed on the device). The mechatronic equipment described in this paper is used to develop neuromuscular control of the human subject having just one leg in contact with the ground.

Paper [4] provides an introduction into the inertial motion analysis field, focusing its attention on the analysis that is performed using modern mechatronic inertial motion capture systems, highlighting both the advantages and drawbacks of using such a system and outlining the main constituent elements of these systems as well as the necessary steps to be carried out in order to be able to accomplish such analysis.

Paper [5] presents the same original solutions for tactile sensor interface topology obtained using PSPICE simulation for design optimisation of tactile resistive and capacitive sensors interfaces.

Patent [6] and patent application [7] describe thoroughly both scientific and practical approaches of designing an experimental model of a complex mechatronic system for human body motion monitoring and analysis.

2 Conceptual Model of the Mechatronic Equipment

2.1 General Specifications

The architecture of the mechatronic equipment includes three distinct subassemblies: a portable electronic equipment called "sensorial", an automation panel and a mechanical subsystem (Fig. 1).

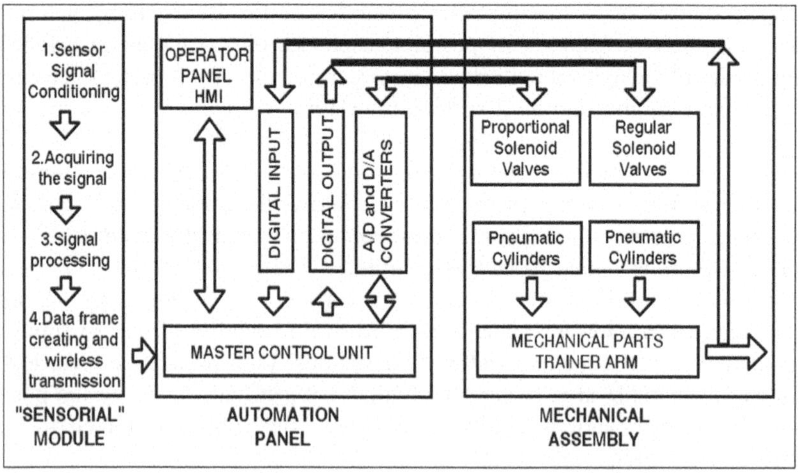

Fig. 1. Architecture of the mechatronic equipment

A set of sensors are placed on the insole or sole of the supporting leg's shoe (the foot that is in contact with the ground), of the human subject under analysis and connected to the electronic sensorial module. The entire mechatronic system is designed in such a way that, depending on how the sensors, which make up the aforementioned system, are stimulated, the command application software will send, to the pneumatic actuator system of the mechanical subassembly, those commands that will put the human subject, in a continuous state of postural adjustment due to his body's constant need to find his balance.

The mechatronic system could implement the same motion cycle, with the same motion parameters, regardless of the support leg used. At the same time, it allows the implementation of individualized motion cycles, whose motion parameters are adapted to each individual's specific biomechanics.

Data Flow between the electronic modules ensures the dynamic configuration of the motion cycle that the human subject will perform.

2.2 Description of the "Sensorial" Module

The electronic module called "sensorial" acquires data from the tactile sensors located on the sole of the support foot of the human subject. The electronic data acquisition module is attached to the support leg of the human subject. The main design requirements of this module are the portability, the wireless data transmission and the power autonomy during each working session.

The movements made by the human subject for rebalancing, as a natural biofeedback reaction, should not in any way be restricted by any electrical connection cables (to prevent stumbling). These considerations require the electronic module to perform data transmission wirelessly in real time. The data transmitted by the "sensorial" portable module are received and made available to the MCU (Master Control Unit) electronic module contained within automation panel.

2.3 Description of the Automation Panel

The automation panel should contain the following hardware components:

1) an electronic interface module for receiving data from the "sensorial" module
2) a Master Control Unit MCU
3) digital inputs modules (used for proximity sensors and operator buttons)
4) digital output modules (used for solenoid valves and indicator lamps)
5) A/D and D/A converters for the movement of the proportional solenoid valves
6) a Control Panel for human-machine interaction.

2.4 Description of the Mechanical Assembly

The mechanical assembly is designed in the form of a manipulator, whose effector element, called the trainer-arm, is actuated by means of a pneumatic driving system. The pneumatic system is designed to execute the commands issued by MCU corresponding to the motion cycle. The position of each cylinder is defined by its associated inductive sensors (up, down and/or middle).

The trainer-arm supports and fastens the human subject's manipulated foot, has the role of imposing on the user suspended leg, a certain trajectory of movement, according to a specific motion cycle, forcing him/her to adapt his/hers posture continuously during moving cycle.

By constantly changing the motion direction, based on a 3D predefined trajectory, of the trainer arm, the person under analysis will be subjected to a continuous state of imbalance. The human subject will have to counteract this continuous state of imbalance, by adapting his posture accordingly, with the intention of returning to a state of equilibrium.

An example of motion cycle is shown in Fig. 2.

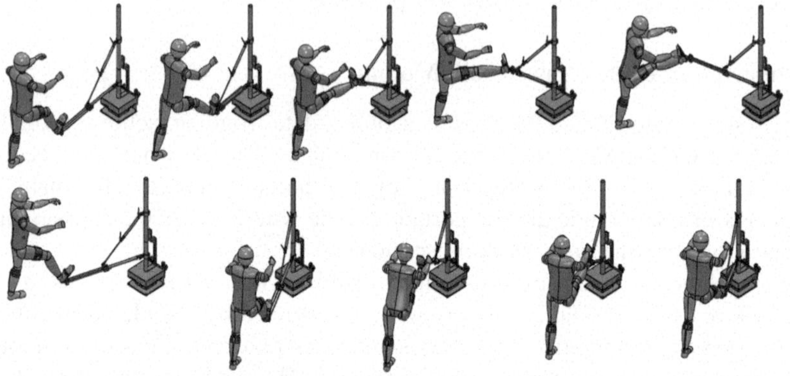

Fig. 2. Motion cycle

Apart from the lower limbs muscles endurance development, this complex mechatronic system can be used for the improvement of coxo-femoral joints mobility (for the athletes). In this case, by using of a remote, the user can control the vertical movement, of his manipulated leg, by commanding the vertical rise of the trainer-arm, which, in time, leads to an increase of the angle between the human subject's lower limbs, both on his sagittal and front planes. In this situation, there is no need for the sensorial module because all the control is held by the user, the only one who can decide how much of an angle increase can he withstand in one session. This is due to the fact that instead of an improvement in the coxo-femoral joints mobility, an improper movement of the trainer-arm can lead to serious accidents.

The action to improve the mobility of the coxo-femoral joints will be achieved by limiting the movement of the trainer-arm, to the vertical plane, in the ascending direction.

3 Proof of Concept

3.1 Hardware and Software Implementation

The "sensorial" module is built around an Arduino Nano board [8] which contains a 8-bit AVR RISC-based microcontroller, which is connected to a transceiver module nRF24L01 which operates in 2.4 MHz ISM frequency [9] as data transmitter.

The complete set of tactile force sensors is not yet available, so for the testing just one tactile force sensor is used [10]. The signal conditioning, from tactile force sensor, is made by a resistor connected to GND. This approach is proven to be sufficient for this proof of concept. Firmware changes have been made to replace the missing sensors as follows: the average of successive values acquired from just one force sensor available is successively compared to several ascending threshold values. The data frame is continuously transmitted and contains the sensor's id associated with the preset threshold value.

The firmware for the "sensorial" module is developed using Arduino IDE 1.8.7. Version.

The "sensorial" module is shown in Fig. 3.

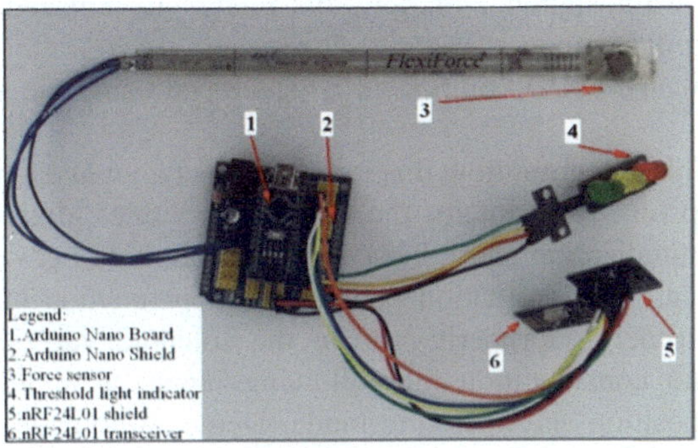

Fig. 3. The "sensorial" module

The automation panel is built around an Arduino Mega 2560 board [11, 12] which contains a powerful 8-bit AVR RISC-based microcontroller and act as a MCU.

The schema of automation panel is shown in Fig. 4.

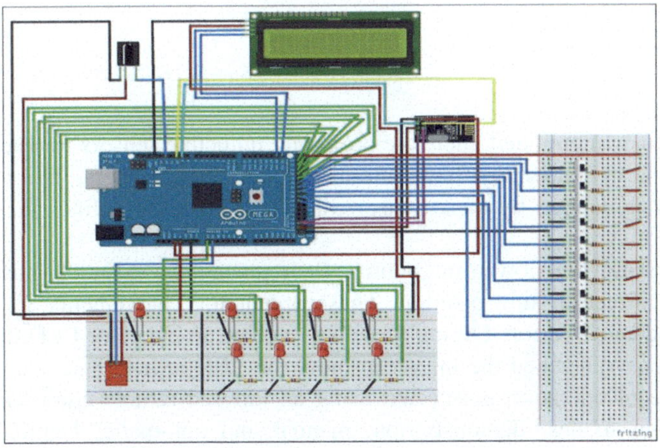

Fig. 4. Schema of the automation panel

The input part of the control panel for human – machine interaction is built using the following components: one TTP223 capacity sensor suitable for touch button and a remote together with its infrared receiver for issuing commands. The output part of the control panel consists one LED that indicates the positioning status of the pneumatic movement cylinders and an I2C 2 × 16 LCD for text messages.

Ten digital inputs of the Arduino Mega Board have to drive the inductive sensors placed in appropriate positions of the pneumatic cylinders, using 24 V to 5 V level translation. For now, the inductive sensors are substituted by two-state buttons connected by 10K pull-up resistors to the digital inputs. A number of eight digital outputs of the Arduino Mega Board have to be connected to 24 V relays, using 5 V to 24 V optical isolated level translation. For now, the relays are replaced with LED diodes connected to the digital outputs using 470-Ω resistors.

The firmware for the MCU component is developed using the same Arduino IDE 1.8.7. Version.

Every Arduino-like program flow diagram contains two mandatory functions: setup and loop. The setup function contains the following routines: digital input/output initializations, serial initialization and radio connection protocol between the "sensorial" electronic module and the automation panel. The loop function implements a Finite State Machine, in order to properly actuate the solenoid valves assigned to each cylinder, taking into consideration the following inputs: data received from radio frames, cylinders' position and operating mode (Demo, Automatic and Manual).

The simplified loop function is the following:

```
void loop() {
   Get_RadioDataLED();
   Get_Digital_Inputs();
   Get_Infrared_Input();
   Get_HMI_Input();
   Finite_State_Machine();
   Set_DigitalOutput();
   Display_HMI_data();
}
```

As previously mentioned, there are proximity inductive sensors for each pneumatic cylinder. Some cylinders also have intermediate position sensors. The movement of each cylinder is defined by these sensors. The motion cycle defines the direction of movement.

After pressing the power on button, each pneumatic cylinder must be moved towards the correct starting position. These preparatory movements are initiated by pressing a dedicated button. During these movements an indicator LED flashes. When the cylinders have reached the initial position, the LED remains lit.

The mode of operation is selected using an infrared remote controller. There are three operating modes: demonstration, manual and automatic. Demonstration and manual modes are used for positioning proximity sensors on cylinders and preliminary testing. Radio data is not taken into account in demo mode and manual mode. In manual mode one can apply commands to any solenoid valve.

In automatic mode, the programmed motion cycle is executed until all the cylinders return to the start configuration (initial position).

3.2 Alternative Implementations of the Automation Panel

There at least two different approaches that can be studied for physical implementation of the automation panel.

The first approach involves using an industrial PC equipped with internal boards having +24 V digital I/O, AD and DA converters. Programming the device would be accomplished using programming languages like C++, C#, Visual Basic or development environments for a visual programming language such as LabView.

The second approach involves using several PLC modules and a dedicated HMI. The authors of this paper have a good knowledge in designing, programming and commissioning industrial equipments made up of Siemens PLC modules from S7-300 family. Digital inputs and digital outputs of the PLC modules operates at +24 V.

For that purpose, a board containing an Arduino Nano board, an nRF24L01 module and a proprietary board having optical couplers has been made. The optical couplers translate the +5 V output of the Arduino Nano pins into +24 V signals.

The Arduino board has been programmed with a minimal piece of firmware that receives radio frames, decodes the data and activates the corresponding digital output.

The module is shown in Fig. 5.

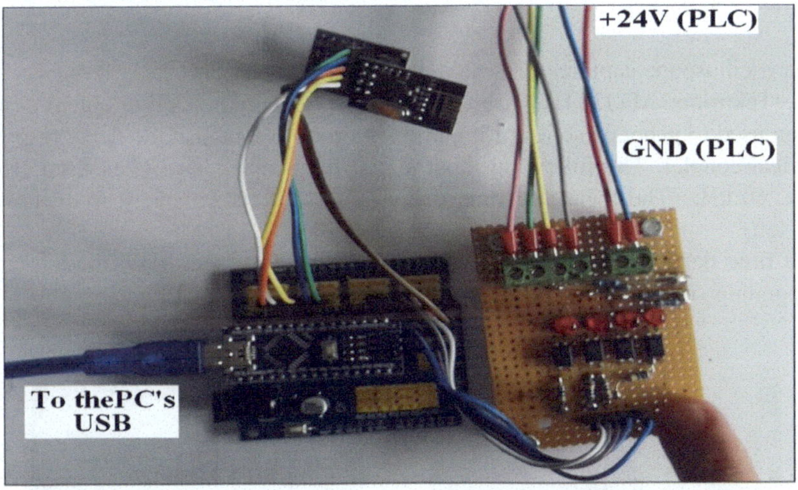

Fig. 5. Data signal adapter board (including radio data receiver)

The main advantages of the alternative approaches are the long-term duty cycle of the electronic modules and the software libraries that may be used into final firmware component. The disadvantages are the high prices of the electronic modules.

4 Results

4.1 Testing Data Transmission and Receiving

Data received by the Arduino Mega 2560 board have been redirected to a PC using serial communication (parameters: 128000, 8, n, 0). Data is received in a Microsoft Excel worksheet in order to evaluate the trigger reference value for biofeedback event.

The graphical representation of a random sequence of values obtained by pressing the tactile force sensor as initial test is shown in Fig. 6.

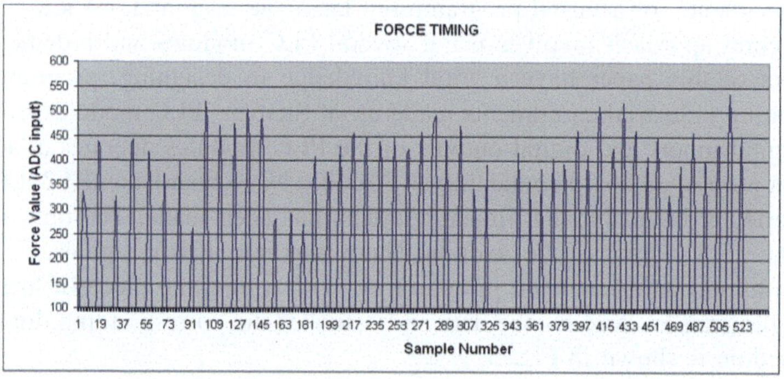

Fig. 6. Graphical representation of the received data contained in radio frames

The oscilloscope captures of time delays are shown in Fig. 7.

The Tektronix AFG3021C Arbitrary/Function Generator Digital signal generator has been used to generate series of impulses in order to establish the delay response of the optical coupler. The following input signal frequencies have been used: 15 kHz, 30 kHz, 60 kHz. The measured time delay is about 8 μs for every frequency. (See Fig. 7, left).

The time delay between output signal issued by the "sensorial" module and the corresponding output channel adapter board is about 1 ms (See Fig. 7, right).

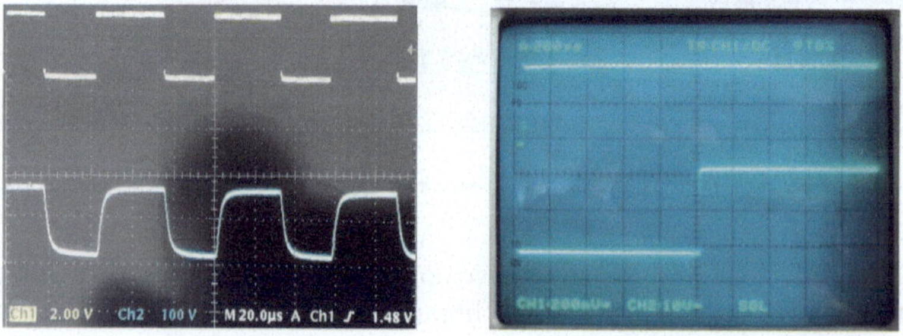

Fig. 7. Time delays. Left: delay introduced by optical couplers. Right: delay between output channel of "sensorial" module and corresponding output channel of the optical coupler

5 Conclusions

The paper presents a conceptual model and a proof of concept for a complex mechatronic system that can be used in neuromuscular control training. It represents a reference point for the development of future complex mechatronic systems that can be used in the field of sports, biomechanics, medicine, physical therapy.

References

1. Freyler, K., Gollhofer, A., Colin, R., Brüderlin, U., Ritzmann, R.: Reactive balance control in response to perturbation in unilateral stance: interaction effects of direction, displacement and velocity on compensatory neuromuscular and kinematic response. https://journals.plos.org/plosone/article?id=10.1371/journal.pone.0144529. Accessed 8 May 2020
2. Serban, I.: Studii şi cercetări privind influenţa mediului înconjurător asupra stabilităţii şi locomoţiei umane, pp. 141–142. http://www.theses.fr/2011ARTO0205/abes/Th%C3%A8se_-_Ionel_-_SERBAN_-_version_-_roumaine.pdf. Accessed 8 May 2020
3. Baritz, M., Cristea, L., Rosca, I.: Dezvoltarea unui sistem de analiză bio-comportamentală a factorului uman în raport cu stimuli externi. Managementul interactiunilor dintre factorul uman si mediul de actiune în vederea optimizării conexiunilor active si pasive. http://old.unitbv.ro/Portals/31/Burse%20doctorale/134378/Seminar/S-04-Baritz%20Cristea%20Rosca.pdf. Accessed 8 May 2020
4. Badea, C.R.: Researches on inertial mechatronic motion analysis systems, based on MEMS. Sci. Bull. Valahia Univ. Mater. Mech. **16**(15), 44–50 (2018). ISSN 1844-1076
5. Constantin, A.: Optimal design and modelling of tactile resistive and capacitive sensors interfaces used in modern mechatronics. ROMJIST-Rom. J. Inf. **20**(4), 400–414 (2017). ISSN 1453-8245
6. Badea, C.R.: Device used for maintaining and/or improving the mobility of coxo-femoral joints, Patent number RO 125003 B1, published on 30 August 2012
7. Badea, C.R.: Device for the development of neuromuscular control/dynamic and static balance, of the strength and endurance of the lower limbs and of the mobility of the coxo-femoral joints, of the athletes. Patents application number A/00889 - 31.10.2017, published on RO-BOPI 4/2019 - 30.04.2019
8. Arduino Nano Homepage. https://store.arduino.cc/arduino-nano. Accessed 2 May 2020
9. NRF24L01 Homepage. https://www.nordicsemi.com/Products/Low-power-short-range-wireless/nRF24-series. Accessed 2 May 2020
10. Force Sensor Homepage. https://www.tekscan.com/products-solutions/force-sensors/a201. Accessed 5 May 2020
11. Arduino Mega 2650 Homepage. https://store.arduino.cc/arduino-mega-2560-rev3. Accessed 4 May 2020
12. Atmel ATMega 2560 data sheet. http://ww1.microchip.com/downloads/en/DeviceDoc/Atmel-2549-8-bit-AVR-Microcontroller-ATmega640-1280-1281-2560-2561_datasheet.pdf. Accessed 4 May 2020

Mathematical Modeling of Torsional Vibrations in a Gearbox with Faults Using Distributed Parameters and Bond Graphs

Daniel Cordoneanu(✉) and Constantin Nițu

Department of Mechatronics and Precision Mechanics, University "Politehnica" of Bucharest, 313, Splaiul Independenței, 060042 Bucharest, Romania
daniel.cordoneanu@stud.mec.upb.ro

Abstract. Vibrations are important in machinery since they present the current state of the system and can indicate certain events (like the presence of a fault). To use the vibrations as indicators for fault detection, one must get the system's ideal response to certain frequencies. To achieve that, mathematical modelling is of great importance. In this paper it is presented a way to model and simulate the response of a gearbox given an input and the presence of a fault. The mathematical approach is based on the distributed parameter model of the shafts and the simulation is done using Bond graphs which is a great way to integrate systems of different nature (electrical, mechanical) in the same model. This approach has the advantage that it is easy to use and integrate in any model and the simulation of faults can greatly reduce costs needed with experimentation.

Keywords: Torsional vibrations · Bond graphs modeling · Simulation

1 Introduction

Fault diagnosis is an important aspect of today's industrial processes as it is more and more convenient to invest into smart algorithms for monitoring that would allow systems to work efficiently and without downtime. In machinery, vibrations are the most used signals for assessing a system's state. Usually, when there is a fault of any sort, the machine's frequency response changes and this allows monitoring systems to detect and isolate a fault. To create such a monitoring system, there can be two approaches:

- Using systems based on data
- Using systems based on models.

In this article, the focus will be on creating a gearbox mathematical model that can provide data to a monitoring system that can be either based on models or based on data.

Gearboxes are elements that are very useful in rotating machinery especially with low torque/high speed motors. They can increase the torque and reduce the speed based on the transmission ratio that can be calculated from the number of teeth/pitch diameters, etc. In this paper, gear transmission basics will not be discussed, as the aim of the article is to model and simulate the torsional vibrations of a gearbox. In [1] Lees et al.

G. I. Gheorghe (Ed.): ICOMECYME 2020, LNNS 143, pp. 100–112, 2020.
https://doi.org/10.1007/978-3-030-53973-3_11

present a way to model and simulate mesh faults in one meshing pair using state equations built on top of the free body diagrams.

This article presents a different approach based on the frequency modes (eigen functions) of the system which when computed can give the system's response.

2 Mathematical Model

For the mathematical model, a gear will be considered at the end of a shaft as in Fig. 1:

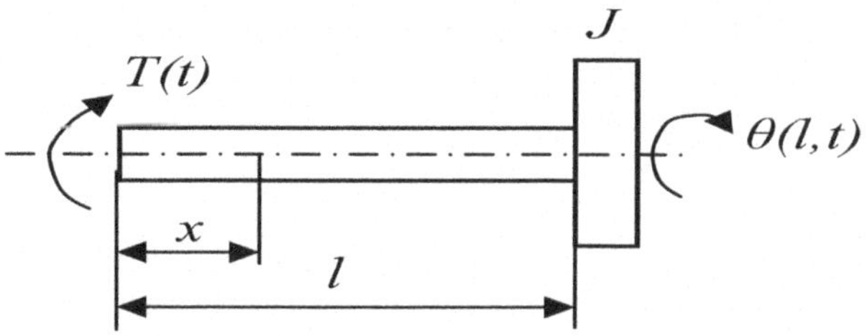

Fig. 1. Shaft and gear representation

Given a shaft of length L and polar moment of inertia Ip has mounted at one of its ends a gear with the moment of inertia J. The torque T(t) is twisting the shaft in the point given by $x = a$. The mathematical model associated to shaft torsion with distributed parameters starts with the equilibrium equation for the strains generated by the forces that act on the faces of a volume element of the shaft, $dV = Adx$, where A is the shaft section area:

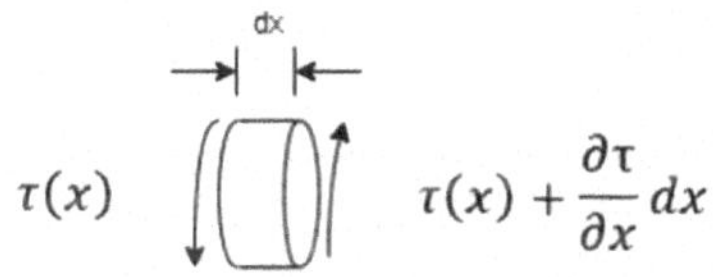

Fig. 2. Stresses on the faces of a volume element

Based on the two figures, Fig. 1 and Fig. 2, the following equation can be deduced:

$$\left[\tau(x,t) + \frac{\partial \tau(x,t)}{\partial x}dx\right] - \tau(x,t) = \rho dx \frac{\partial^2 \theta}{\partial t^2} - \frac{T(t)}{I_p}dx \cdot \delta(x-a) \tag{1}$$

where ρ is the density of the shaft material, $\delta(x)$ is the Dirac function, Ip is the polar moment of inertia of the shaft.

The difference of the tangential tensions on the opposite faces of the cylinder is generated by the inertia forces and the torque applied in $x = 0$. The first term of the right-hand side of Eq. (1) is the result of the shaft's distributed inertia and the second term represents the contribution of the torque *T(t)*. The difference can be computed as:

$$\tau_T = G\gamma_T = G\frac{T(t)dx}{GI_p}\delta(x-a) = \frac{T(t)dx}{I_p}\delta(x-a) \tag{2}$$

Where G is the shear modulus of the shaft's material. Given:

$$\tau(x,t) = G\gamma(x,t) = G\frac{\partial\theta(x,t)}{\partial x} \tag{3}$$

Where $\theta(x, t)$ is the rotation angle of section x at time t, Eq. (2) can be written as:

$$\rho\frac{\partial^2\theta(x,t)}{\partial t^2} = G\frac{\partial^2\theta(x,t)}{\partial x^2} + \frac{T(t)}{I_p}\delta(x-a) \tag{4}$$

The free vibrations are expressed through natural frequencies which can be described using the wave equation [2]:

$$\frac{\partial^2\theta(x,t)}{\partial x^2} = \frac{1}{c^2}\frac{\partial^2\theta(x,t)}{\partial t^2} \tag{5}$$

where $c = \sqrt{\frac{G}{\rho}}$, the speed of the torsional waves. The boundary conditions for a free shaft at both ends are:

$$x = 0, T(t) = 0{:}GI_p\frac{\partial\theta(0,t)}{\partial x} = 0 \tag{6}$$

$$x = l, T(t) = 0{:}GI_p\frac{\partial\theta(l,t)}{\partial x} = 0 \tag{7}$$

The solution is computed using the separation of variables:

$$\theta(x,t) = \overline{\theta}(x)\cdot e^{j\omega t} \tag{8}$$

Equation (8) is introduced into the wave Eq. (5) and the differential equation has the solution:

$$\overline{\theta}(x) = Asinkx + Bcoskx \tag{9}$$

where $k = \omega/c$ is the wave number. The integration constants are computed by using the boundary conditions. Given:

$$\frac{d\overline{\theta}(x)}{dx} = k(Acoskx - Bsinkx) \tag{10}$$

And replacing in the boundary conditions, the constants are:

$$x = 0: A \cdot 1 - B \cdot 0 = 0 => A = 0 \tag{11}$$

$$x = l: 0 \cdot \cos(kl) - B \cdot \sin(kl) = 0 \tag{12}$$

Since the solution with both constants 0 is not of interest, from Eq. (12) we can extract the form of k and implicitly ω:

$$k_n = \frac{n\pi}{l} \quad \text{or} \quad \omega_n = \frac{n\pi}{l}\sqrt{\frac{G}{\rho}} \tag{13}$$

For mathematical convenience, B will be chosen as 1 and the eigenfunctions of the free vibrations have the solution:

$$\overline{\theta}_n(x) = cos\left(\frac{n\pi}{L}x\right) \tag{14}$$

The eigenfunctions have several properties which are worth mentioning, considering beam theory:

1. If a beam is deformed initially under the eigenfunction, it will oscillate harmonically, the oscillation having the form of that eigenfunction mode
2. Any vibration response of the beam can be expressed as a linear combination of the eigenfunctions
3. The eigenfunctions are orthogonal.

Of these three properties, the third is the most important and can be expressed mathematically as:

$$\begin{cases} \int_0^l \overline{\theta}_n(x)\overline{\theta}_p(x)dx = 0, n \neq p \\ \int_0^l cos\left(\frac{n\pi}{L}x\right) \cdot cos\left(\frac{p\pi}{L}x\right)dx = \frac{l}{2}, n = p \end{cases} \tag{15}$$

To find the frequency response of the force vibrations, property number 2 will be used, the response being decomposed as follows:

$$\theta(x,t) = \sum_{n=0}^{\infty} \overline{\theta}_n(x)\beta_n(t) \tag{16}$$

Using Eq. (3) in Eq. (1):

$$\rho\frac{\partial^2\theta(x,t)}{\partial t^2} = G\frac{\partial^2\theta(x,t)}{\partial x^2} + \frac{T(t)}{I_p}\delta(x-a) \tag{17}$$

And substituting the term $\theta(x, t)$ from Eq. (17) in Eq. (18), the following equation is built:

$$\sum_n \rho I_p \overline{\theta}_n \ddot{\beta}_n = \sum_n G I_p \frac{d^2\overline{\theta}_n}{dx^2} \beta_n + T(t) \cdot \delta(x - a) \tag{18}$$

Equation (18) is multiplied with the eigenfunction for mode m, $\overline{\theta}_p(x)$ and both sides are integrated for the length of the shaft:

$$\sum_{n=1}^{\infty} \left(\int_0^l \rho I_p \overline{\theta}_n \overline{\theta}_m dx \right) \ddot{\beta}_n = \sum_{n=1}^{\infty} \left(\int_0^l G I_p \frac{d^2\overline{\theta}_n}{dx^2} \overline{\theta}_m dx \right) \beta_n + \int_0^l T(t)\delta(x - a)\overline{\theta}_m(x)dx \tag{19}$$

In Eq. (19), the second derivative of an eigenfunction can be computed as:

$$\frac{d^2\overline{\theta}_n}{dx^2} = -\frac{\rho}{G}\omega_n^2 \overline{\theta}_n \tag{20}$$

Also, in Eq. (19) every term of the sum is equal to 0 unless $m = n$:

$$\left(\int_0^l \rho I_p \overline{\theta}_n^2 dx \right) \ddot{\beta}_n + \left(\int_0^l \rho I_p \overline{\theta}_n^2 dx \right) \omega_n^2 \beta_n = \int_0^l T(t)\delta(x - a)\overline{\theta}_n(x)dx \tag{21}$$

Given the equation of motion for undamped linear systems, from (21) the modal masses can be extracted as:

$$m_n = \int_0^l \rho I_p \overline{\theta}_n^2 dx = \rho I_p \int_0^l \overline{\theta}_n^2 dx = \frac{\rho I_p l}{2} = \frac{\rho \pi r^4 l}{4} = \frac{m r^2}{4} = \frac{J_s}{2} \tag{22}$$

Where r is the shaft's radius and J_s is the shaft's moment of inertia.

Mode $n = 0$, characterized by $\omega_0 = 0$, represents the rigid movement of the shaft. The modal mass for this mode can be computed as follows:

$$m_0 = \int_0^l \rho I_p \overline{\theta}_0^2 dx = \rho I_p \int_0^l dx = \rho I_p l = \frac{\rho \pi r^4 l}{2} = \frac{m r^2}{2} = J_s \tag{23}$$

The modal stiffness can be also computed based on (22):

$$k_n = m_n \omega_n^2 \tag{24}$$

The right-hand side of Eq. (21) is called the modal effort function:

$$\int_0^l T(t)\delta(x - a)\overline{\theta}_n(x)dx = T(t)\overline{\theta}_n(a) \int_0^l \delta(x)dx = T(t)\overline{\theta}_n(a) \tag{25}$$

Basically, the eigenfunctions and the natural frequencies are determined using free vibrations. Based on (22), (23), (24) and (25), the motion equation is:

$$m_n \ddot{\beta}_n + k_n \beta_n = T(t)\overline{\theta}_n(a) \tag{26}$$

And the frequency response is represented through Eq. (16).

3 Bond Graphs Simulation

3.1 Summary of Bond Graphs

The Bond graphs (BG) represent a modeling language which can connect system components from different power fields. Modeling with BG is based on physical laws of the power flow between elements of a system. These ones are described by equations relating energy-bound variables called effort (force, torque, voltage) and flow (velocity, angular velocity, current). In Fig. 3, the transfer of power is represented to/from an element.

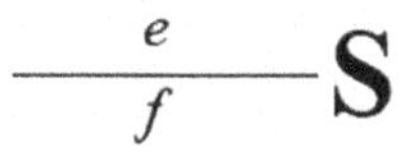

Fig. 3. Power transfer from/to an element

There are three major elements defining element properties in a Bond graph:

- inertial elements that have the symbol **I** (for example a mass or a moment of inertia)
- compliant elements that have the symbol **C** (for example stiffness or compliance)
- resistive elements that have the symbol **R** (for example damping coefficient).

All these elements are linked together by transferring power from a source that go through junctions. The junctions can be of type **0** (the same effort for all the nodes, only one node can introduce effort in this junction) or of type **1** (the same flow for all the nodes, only one node can introduce flow in this junction).

There are two ways in which the power variables can be transformed, while conserving power:

- using a transformer which has the symbol **TF**
- using a gyrator which has the symbol **GY**.

Further in the paper, the simulation will be done using Bond graphs and using the modelling program 20-sim.

3.2 Approximation of the Frequency Response Based on Eigenfunctions

As the frequency response of a beam (shaft in our example) can be composed from eigenfunctions as in Eq. (16), to use the Bond graphs the following variables will be formed:

$$p_n = m_n \dot{\beta}_n = I_n \dot{\beta}_n \tag{27}$$

$$q_n = \beta_n \tag{28}$$

Using Eqs. (26), (27) and (28) the following state equations are obtained:

$$\frac{dp_n}{dt} = -q_n k_n + T(t)\overline{\theta}_n(a) \tag{29}$$

$$\frac{dq_n}{dt} = \frac{p_n}{I_n} \tag{30}$$

In the associated Bond graph, the modal Eqs. (29) and (30) are represented for every $n = 0, 1, 2, \ldots$ All these modes are getting energy from the exterior torque *T(t)* and the causality is integral for all the components. The transformers will have the multiplication factor equal to $\overline{\theta}_n(a)$. For the rigid mode $n = 0$, there is no stiffness. In Fig. 4 an example for a simple beam is presented.

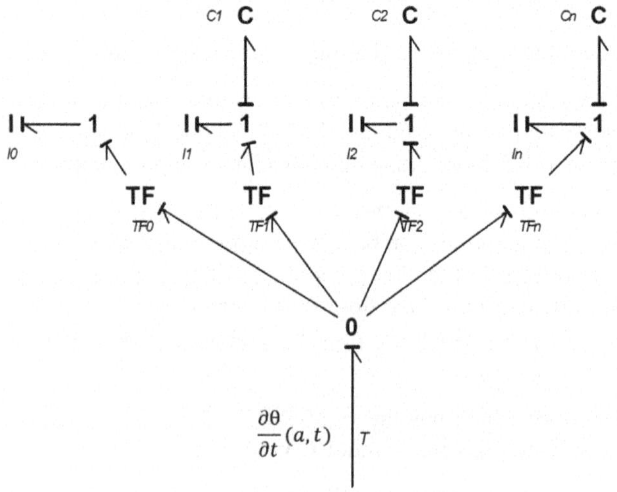

Fig. 4. Bond graph for a simple beam

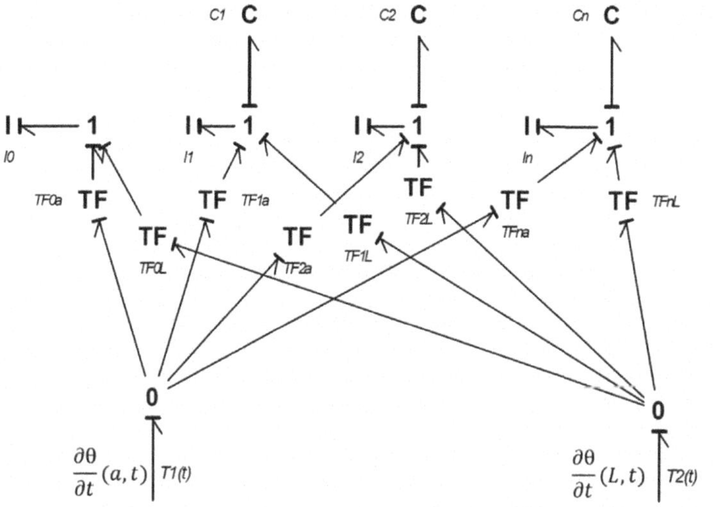

Fig. 5. Bond graph for a beam upon which two torques act

If there is a second torque on the beam in the point $x = L$, the Bond graph changes to the one in Fig. 5.

3.3 Frequency Response for a Gear Mounted on a Shaft

A shaft behaves like a beam and a gear mounted can be considered as a second torque that acts on the point $x = L$ where L is the length of the shaft. Given the system, the Bond graph presented in Fig. 5 will have an additional junction of type 1 for the gear. This junction will have the inertial element as the moment of inertia of the gear and a small compliance element so that the gear has a rigid movement.

If $T_1(t)$ acts on $x = 0$, then the transformers coefficients will be:

$$\overline{\theta}_n(0) = cos\left(n\pi\frac{0}{L}\right) = 1 \tag{31}$$

Also, for the torque $T_2(t)$ acting on $x = L$, the coefficients will be:

$$\overline{\theta}_n(L) = cos\left(n\pi\frac{L}{L}\right) = (-1)^n \tag{32}$$

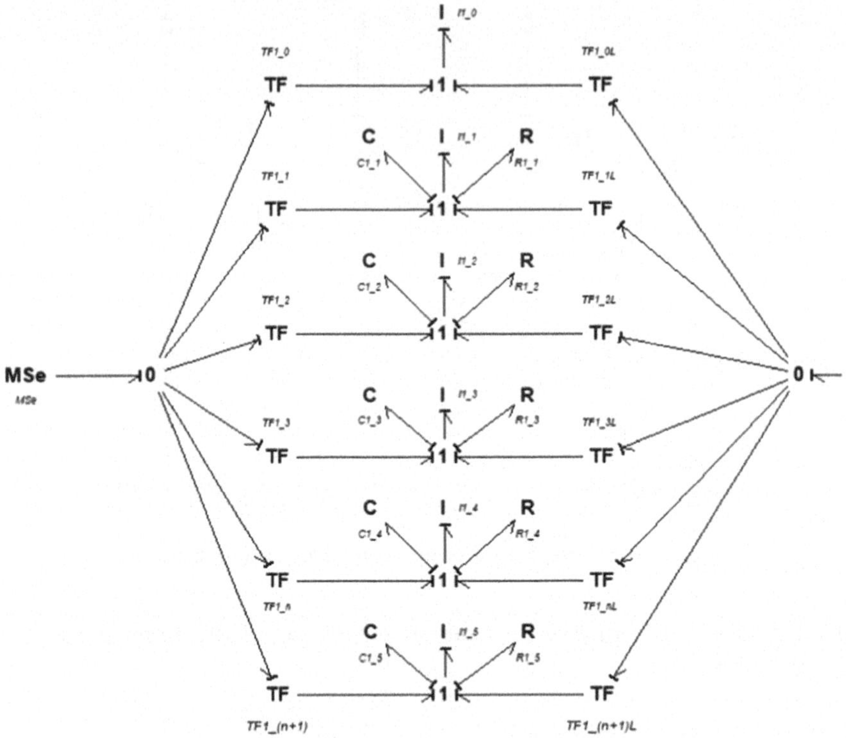

Fig. 6. Bond graph for driving gear of a pair of gear mesh

Since the aim of this paper is to simulate the meshing of two gears, in Fig. 6, Fig. 7 and Fig. 8 an example Bond graph is given for this purpose, also having the stiffness and damping of the mesh included and a modulated flow source in order to introduce teeth faults. The three figures are part of the same model but since the model is too big it could not be included in only one figure.

Damping should also be added for a better simulation. According to [3], damping should not be modeled in detail, but added for each mode as a resistive element with the following value:

$$R_n = 2\xi_n m_n \omega_n \tag{33}$$

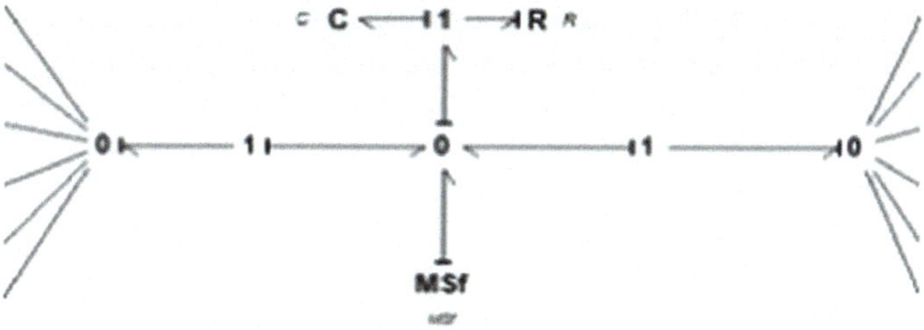

Fig. 7. Bond graph for the meshing part of a pair of gear mesh

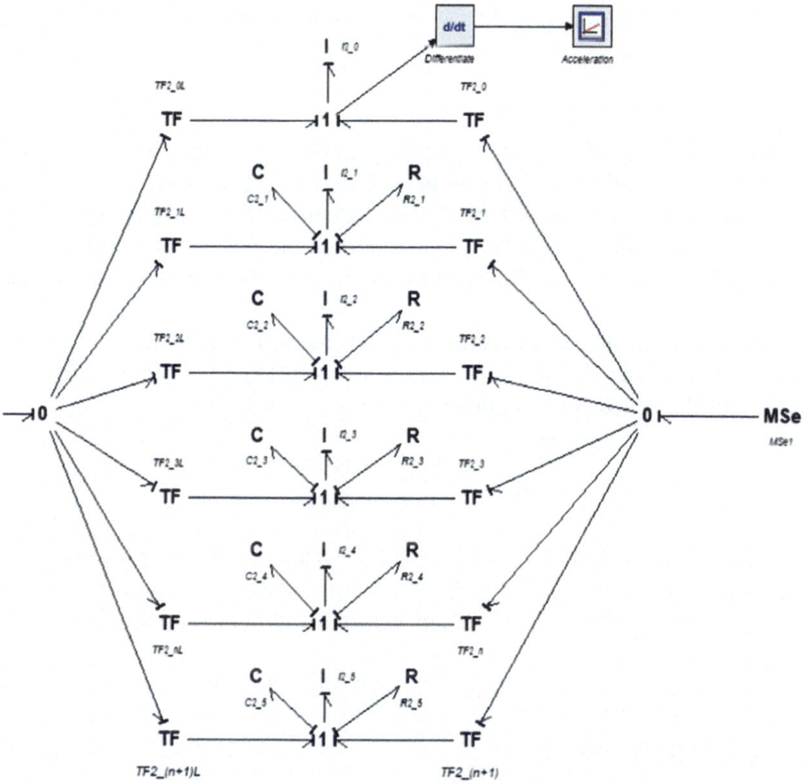

Fig. 8. Bond graph for the driven gear of a pair of gear mesh

4 Simulation and Results

For the simulation, two spur gears were created in Solidworks to get the inertial properties directly out of the CAD program. The two gears are identical meaning that the transmission ratio is 1. This is only for simulation purposes and checking the behavior of an induced flow on the frequency response of the mesh.

According to Eq. (17) all the modes need to be computed up to a point. To compute each mode value, the properties of the shaft and gears are needed. In this example, the material used is carbon steel. The properties of the shaft and gear are presented in Table 1.

Table 1. Gear and shaft properties

	Gear	Shaft
Length [m]	0.02	0.2
Shear modulus (G) [Pa]	80e9	80e9
Moment of inertia [kg m^2]	3.2e−3	930e−6
Density [kg/m^3]	7858	7858
Radius [m]	0.07	0.01

Obviously, not all modes will be computed. In this paper, the criteria for getting the number of nodes will be based on numerical reasons. This way, from Eq. (25) the stiffness exhibits quadratic growth with the frequency. The maximum mode will be chosen by when the stiffness will be 100 higher than the first computed stiffness, as the higher the stiffness value is, the less contribution the mode will have on the frequency response.

Besides computing the mode parameters (stiffness, damping factor), the meshing elements must be computed as well. According to [4] (p. 146), the stiffness can be computed with the following equation:

$$k_m = \frac{0.25 d_w^2 b E_1 E_2 Z_1 Z_2}{E_1 Z_1 + E_2 Z_2} \tag{34}$$

where: d_w is the pitch diameter, E_1, E_2 - Young modules of the gear materials, Z_1, Z_2 - elastic deformation factors for the gears, b - length of the teeth contact. The damping factor for the mesh can be also computed with the following equation [5]:

$$c_m = 2\xi_m \sqrt{k_m m_e}$$

where m_e can be computed as:

$$m_e = \frac{I_1 I_2}{I_1 R_2^2 + I_2 R_1^2} \tag{35}$$

Where I_1, I_2 are the moment of inertias of the gears and R_1, R_2 are the pitch radii of the gears.

If the fault is introduced as flow inside the system having the form of a wave with the period equal to ω_m then the amplitude will be proportional to the size of the fault (tooth crack). In the next two figures, the angular displacement is presented when there is no fault and with a fault introduced using a waveform with amplitude 0.05.

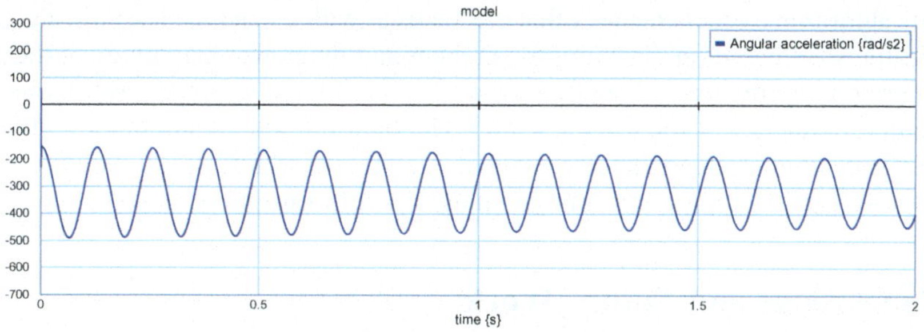

Fig. 9. Angular acceleration when there is no fault

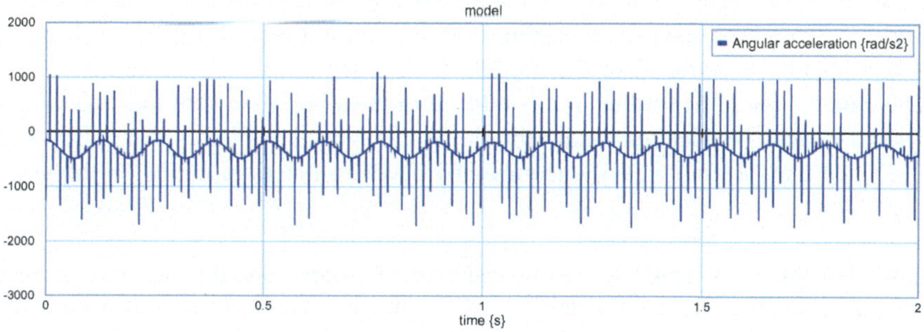

Fig. 10. Angular acceleration for the gear mesh with the fault present

In Fig. 11, the data presented in Figs. 9 and 10 is processed with the algorithm described in [6].

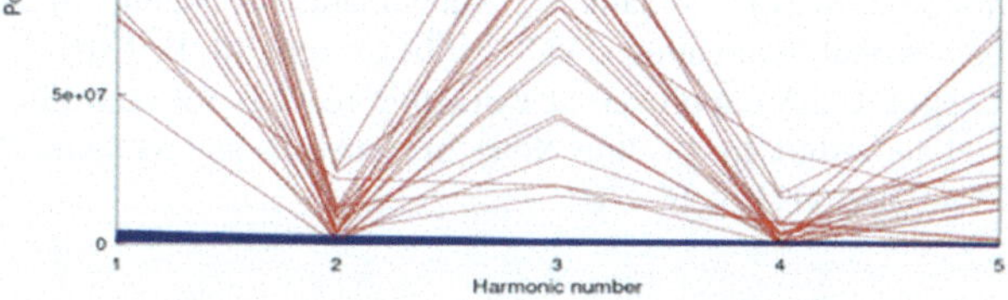

Fig. 11. Processed data for the gear mesh with no fault and with the fault present

This algorithm extracts features that can be used with a machine learning algorithm; these features are based on the harmonic number and the power magnitude of the signal at the harmonic's frequency. The first harmonics (and the monitored base frequency is the rotation frequency of the driving shaft). The fault introduces amplitudes which can be clearly distinguished in the represented features, therefore these features can be successfully used to train a machine learning algorithm for fault diagnosis.

5 Conclusions and Future Work

As it can be seen from Fig. 9, Fig. 10 and Fig. 11, as the amplitude of the fault waveform increases, the frequency response gets more and more distorted and spikes can be seen appearing from the initial ideal response. In conclusion, the simulation can be used for generating data for modeling torsional vibrations of a gear mesh pair and faults can also be introduced in the model.

The data can be used further in a condition monitoring algorithm that would assess the system's state. Besides torsional vibrations, transversal oscillations are also of great interest since they are more easily acquired by accelerometers mounted on the gearbox case.

Acknowledgment. This work has been funded by the European Social Fund from the Sectoral Operational Programme Human Capital 2014–2020, through the Financial Agreement with the title "Scholarships for entrepreneurial education among doctoral students and postdoctoral researchers (Be Antreprenor!)", Contract no. 51680/09.07.2019 - SMIS code: 124539.

References

1. Lees, A.W., Friswell, M.I., Litak, G.: Torsional vibration of machines with gear errors. In: Journal of Physics: Conference Series (2011)
2. Beards, C.F.: The vibration of systems with distributed mass and elasticity. In: Beards, C.F. (ed.) Engineering Vibration Analysis with Application to Control Systems. Elsevier, London (1995)
3. Karnopp, D.C., Margolis, D.L., Rosenberg, R.C.: System Dynamics: Modeling, Simulation, and Control of Mechatronic Systems, 5th edn. John Wiley & Sons, Hoboken (2012)
4. Tudor, D., Nitu, C., Demian, T., Curita, I.: Bazele proiectarii aparatelor de mecanica fina, vol. 2. Editura Tehnica, Bucuresti (1986)
5. Omar, F.K., Moustafa, K.A.F., Emam, S.: Mathematical modeling of gearbox including defects with experimental verification. JVC J. Vib. Control **18**(9), 1310–1321 (2012)
6. Cordoneanu, D., Niţu, C.: An approach of extracting features for fault diagnosis in bearings using the Goertzel algorithm. In: Lecture Notes in Networks and Systems (2020). https://doi.org/10.1007/978-3-030-26991-3_16

An Approach on Predicting a Machine's Effector Vibrations Based on Motor Vibrations Using a Regression Artificial Neural Network

Daniel Cordoneanu(✉)

Department of Mechatronics and Precision Mechanics, University "Politehnica" of Bucharest, Splaiul Independenței 313, 060042 Bucharest, Romania
daniel.cordoneanu@stud.mec.upb.ro

Abstract. Vibrations are mechanical movements which can offer valuable information about the state of a machine. Being able to simulate and predict this type of signal is of great importance in fault diagnosis and condition monitoring of a machine since vibrations are considered the most used and most efficient way to detect and diagnose a fault. Since simulating vibrations can be a difficult task given the non-linear equations that need to be solved, machine learning algorithms can offer a great solution to extract information from data and predict an outcome close to reality given an input signal. Obviously, the input signal is processed, normalized, and fetched to the neural network. In this paper, a regression neural network is used for predicting the vibrations of an effector of a one-dimensional system which moves in only one direction. The vibrations are measured in the direction of the effector's movement while the vibrations of the motor are acquired by the motor's support.

Keywords: Vibrations · Machine learning · Fault diagnosis

1 Introduction

Fault diagnosis has been getting more and more important for the industrial field as with the modern hardware and software technologies it is possible to prevent a lot of faults and save a lot of downtime and manpower. Fault diagnosis is mainly composed of fault detection and fault isolation. The first part monitors the presence of a fault while the second tries to pinpoint the location and severity of the fault.

Fault diagnosis methods can be split up into two main categories:

- Model based fault diagnosis – the model of the system is created where a decision is made based on the residual computed from the signal acquired from the real system and the modelled system,
- Data based fault diagnosis – data driven models that represent the real system based on acquired data from sensors.

Since processes tend to get more and more complex, also the non-linearities that come up in the models are more difficult to solve numerically and more prone to errors.

G. I. Gheorghe (Ed.): ICOMECYME 2020, LNNS 143, pp. 113–122, 2020.
https://doi.org/10.1007/978-3-030-53973-3_12

Given this, data-driven models gain a lot of attention as machine learning models are under extensive development.

There are many types of machine learning models and they can be split under two great categories:

- Supervised learning models – models that need labels for output data to create a model between the inputs and outputs – very useful when pre-recorded data is available
- Unsupervised learning models – they create their own representation and group the data by certain measures set by the model's designer – very useful when new data needs to be grouped and classified in a new class.

Each model has its own advantages and disadvantages in different areas. For instance, for fault detection, an unsupervised model can be used. In [1], Costa et al. present a new evolving classifier algorithm that is based on recursive density estimation. They use this approach for fault detection and isolation in an industrial plant with success by first finding outliers in the data and then classifying them.

Also, for fault detection a neural network can be used (which is on of the many supervised learning models) to classify faults. For example, Silvestri et al. used a neural network in [2] for detecting faults in sensors in a flight control system without sensor redundancy. They use the Extended Backpropagation algorithm which is a new method which offers improvements from the normal backpropagation algorithm.

Further in this article, an approach for using a regression neural network to fit input data coming from motor vibration to output data coming from the effector vibrations. The data will be processed first by extracting mel-frequency cepstral coefficients (MFCC) and then fetched to the neural network. The neural network will be trained using backpropagation together with ADAM method and will be tested on two batches: cross validation data and testing data.

2 Processing the Data into MFCC

This article will not go into details about the mel-frequency cepstral coefficients (MFCC) but will offer a small background for contiguity reasons.

MFCC are especially used in speech recognition to extract useful features from the speech acoustic signal. These coefficients are particularly useful because they identify parts of the audio signal for classifying the spectral form. They are extracted based on energy filters which have a good resolution in the low frequency spectrum, where fault frequencies usually appear. For instance, in [3] Nelwamondo and Marwala used MFCCs to successfully diagnose bearing faults. Also, Singh and Meena used these coefficients for fault diagnosis of car engines [4].

The algorithm to extract MFCCs is presented in the following steps:

1. The original signal is split up into short frames
2. For each frame, the spectral density is computed
3. For each frame, the mel-filter banks are applied and energies for each filter is summed up

4. The energies obtained in step 3 are taken to a log scale
5. The discrete cosine transform (DCT) is applied on the energies obtained at step 4
6. The coefficients from the transform are used starting with coefficient 2 (in speech recognition goes up to 13, but in this case, higher frequencies may be of interest as well).

The above steps are needed to have the signal analyzed, filtered, and normalized properly. Having short frames means having a stationary signal in each frame. Applying the mel-scale allows the lower frequencies energies to have a better resolution and by taking the energies to a log scale allows normalization and getting the cepstral coefficients. Because the filter banks overlap, the DCT is needed to decorrelate the monitored frequencies.

The mel-scale can be mathematically expressed as [5]:

$$m = 1127\ln\left(1 + \frac{f}{700}\right) \tag{1}$$

Where f is the frequency in Hz.

The inverse mel-scale transformation can be expressed as:

$$f = 700\left(e^{\frac{m}{1127}} - 1\right) \tag{2}$$

In Fig. 1 an example of 30 mel filter banks is shown:

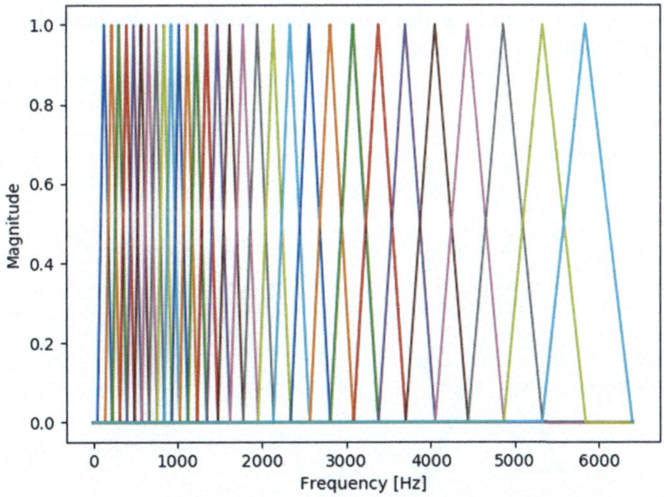

Fig. 1. Thirty mel filter banks

3 Proposed Algorithm

3.1 Neural Networks

The neural network is a machine learning algorithm which is widely used in classifying tasks or predictive tasks. There are numerous types of neural networks, but the focus of this article will be the multi-layer perceptron (MLP) neural network.

This algorithm is part of the supervised learning algorithms. The MLP is used for non-linear fitting models and is composed out of layers:

1. Input layer
2. Hidden layers
3. Output layer

The input layer represents the data that comes into the network. The output layer is composed of the result of the mathematical operations applied on the input layer data. The hidden layers are composed out of units which have a weight and an activation function. The units in two consecutive layers are all connected one to each other (dense neural network).

The activation function is usually is a non-linear function (logistic function, sigmoid function) which allows a unit to pass the data further or not. The weight is the factor which increases or decreases the output of a unit should that unit be active in a layer.

The data is passed from the input layer to the hidden layers and then to the output layer.

To train an MLP, usually backpropagation is used together with a loss function and an optimization method (gradient descent, adam, stochastic gradient descent, etc.). In Fig. 2 an example MLP is presented with one hidden layer.

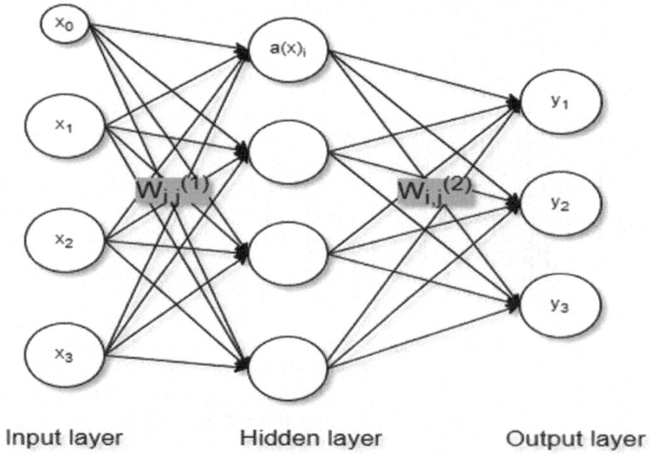

Fig. 2. MLP with one hidden layer

3.2 The Problem Description

The problem described and solved in this article is the mapping of the vibrations coming from a motor to the vibrations coming from the system actuated by that motor. Given A as the space made of all the vectors of MFCC extracted from the vibration data of a motor that actuates the system and given B as the space of all the vectors of MFCC extracted from the vibration data of the moving system, let T be a system so that:

$$\tilde{Y}(t) = X(t)T \tag{3}$$

Where $\tilde{Y}(t)$ is the estimated value of the MFCC given by the T system at time t if the input is $X(t)$. The following conditions must be fulfilled:

$$\begin{cases} X(t) \in A \\ \tilde{Y}(t) \in B \end{cases} \tag{4}$$

If $Y(t)$ represents the MFCC vector at time t acquired from the real sensor on the system, then the T system can successfully map the A space to the B space if:

$$abs\left(\tilde{Y} - Y\right) \leq \varepsilon \tag{5}$$

Where ε is a threshold value that is chosen by trial and error until the prediction is correct.

If the system satisfies the conditions stated in Eqs. 4 and 5 during cross-validation and testing process, then any violation of this equation means that there is a fault in the real hardware.

The T system can be taught of as all the transformations that occur on the vibrations of the motor and get passed to the end effector. If the vibration data processed from the motor is not part of the A space, then a fault occurred in the motor. If

$$abs\left(\tilde{Y} - Y\right) > \varepsilon \tag{6}$$

Then a fault occurred in the kinematic chain that affects the real hardware, but the system that simulates the transmission is given a different response to the input. This approach based on a residual information coming from a data-based model allows non-linear response to be created based on each system's characteristics and mathematical models not being needed to be created for each system specifically.

This algorithm also should be able to detect faults which are not easy to spot only based on the vibration data coming from the motor and from the end effector. For instance, if there is looseness or slip and the motor moves too fast, the end effector will not move. However, both X and Y will be part of the precalculated spaces.

Further in the article the T system can be considered a neural network, but it can be any other system capable of modelling a non-linear behavior and mapping two separate spaces.

3.3 Algorithm Description

The algorithm can be described with the following steps:

1. Place one accelerometer on the motor's plate
2. Place one accelerometer on the end effector so that the vibrations in the direction of movement are acquired
3. Start data acquisition for both accelerometers on using the same acquisition board
4. Create MFCC from the acquired data
5. Split the data into training data, cross-validation data, and testing data by using the following ratios: 70%, 15%, 15%
6. Train the neural network on the training data and check its efficiency using different configurations (multiple layers, different number of units on each layer) until ε is reached
7. Store the threshold value and the neural network model
8. Use the model and the threshold value to detect faults on the real system.

In Fig. 3 the algorithm's diagram is presented, being split into two parts: the left part presenting the training process while the right part presenting the monitoring process that would happen on the real system.

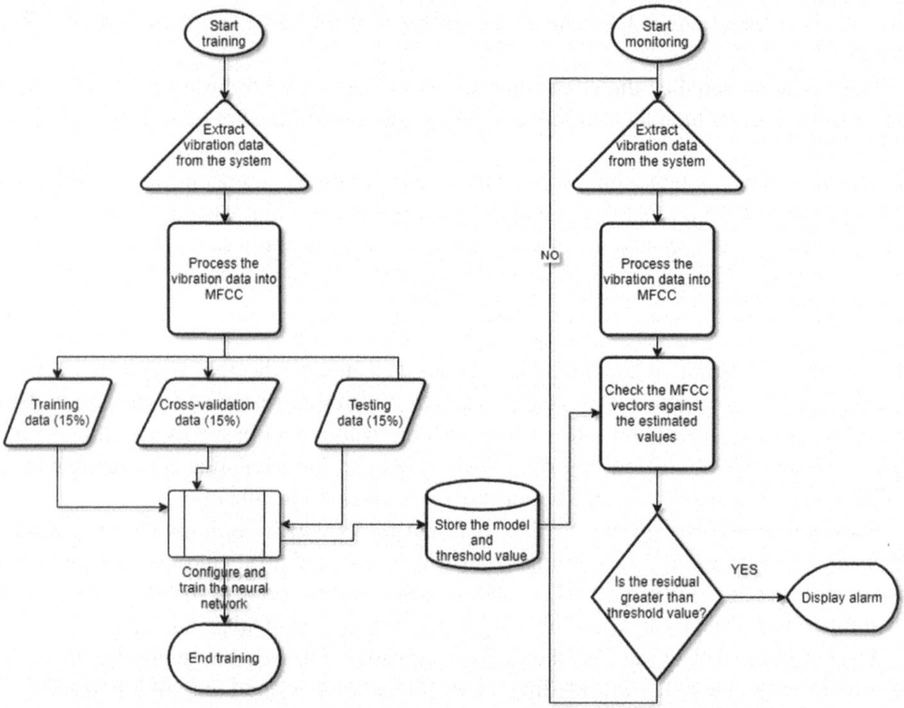

Fig. 3. Algorithm diagram

4 Results and Discussions

Given the system sketched in Fig. 4, two accelerometers were used:

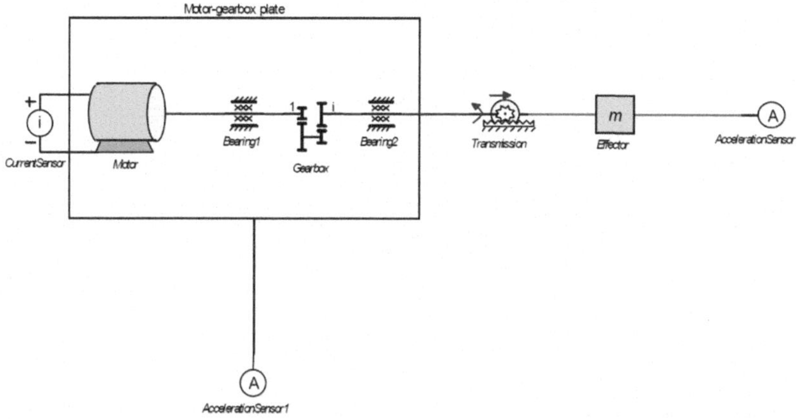

Fig. 4. Sketch of the described system

Both sensors were wired to a data acquisition board from National Instruments (which allows multiple channel acquisition) and a data acquisition program was created in the software provided by National Instruments, LabView.

The maximum sampling rate of the board is 25.6 kHz so that being used for two channels, means 12.8 kHz acquisition sampling rate. Therefore, by Shannon's theorem, the maximum frequency that can be measured before aliasing is 6.4 kHz.

Once the vibration signals have been acquired, they are processed into MFCC using python libraries which are specifically designed for this.

The MFCC are put in matrices $m \times n$ where m represents the number of collected samples and n represents the number of coefficients kept. In this case, 30 coefficients are extracted.

Theoretically, a neural network can fit any non-linear function using only one hidden layer. There are few cases where multiple hidden layers improve the model's accuracy. In this case, the approach will be to use at first 1 hidden layer with up to 120 units and then add a second layer to check if the accuracy of the model improves.

The neural network will be trained using the TensorFlow library. The network will use as optimizer the Adam method described in [6] and the mean squared error loss function. The mean squared error can be computed as:

$$MSE(y, \tilde{y}) = \frac{1}{N}\sum\nolimits_{i=0}^{N} (y_i - \tilde{y}_i)^2 \quad (7)$$

Where $\tilde{y}$ is the predicted value of the model. This loss function is widely used in regression model.

To choose the correct configuration, the R squared statistic will be used for the neural network. This statistic is useful in regression problems because it represents the dependency of the variance of a dependent variable to the independent variables (inputs). If the R squared is 0.7, it means that 70% of the observed variance can be explained by the input's variance.

The R squared can be computed as follows:

$$\bar{y} = \frac{1}{N}\sum\nolimits_{i=1}^{N} y_i \tag{8}$$

$$TDV = \sum\nolimits_{i=1}^{N} (y_i - \bar{y})^2 \tag{9}$$

$$RS = \sum\nolimits_{i=1}^{N} (y_i - \tilde{y}_i)^2 \tag{10}$$

$$R^2 = 1 - \frac{RS}{TDV} \tag{11}$$

Where:

- N is the number of samples
- y is the output data on which the training takes place
- $\bar{y}$ is the mean of the observed data
- $\tilde{y}$ represents the predicted values of the model
- TDV is the total data variance multiplied by N.

By training the neural network with the above-mentioned parameters and extracting the R squared statistic, the following results were achieved:

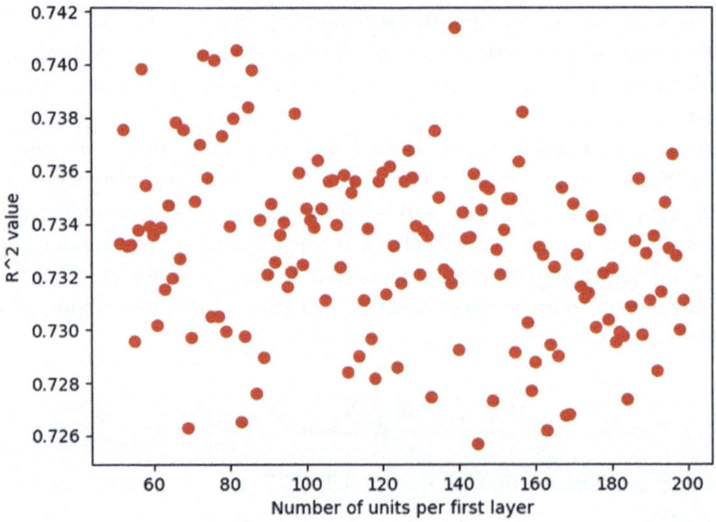

Fig. 5. R^2 statistic for the neural network with one hidden layer

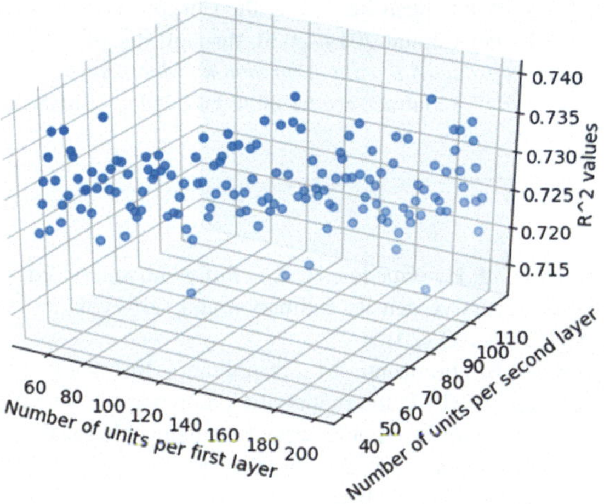

Fig. 6. R^2 statistic for the neural network with 2 hidden layers

As it can be seen from Figs. 5 and 6, the best R^2 value is obtained when the neural network has one hidden layer. When a second layer is added, the neural network's accuracy does not improve, but it decreases a bit.

Based on the above-mentioned figures, the configuration for the neural network has been chosen to have one hidden layer with 80 units.

Further, the model was tested using the test data and using an isolation forest [7] for checking whether the predicted value of the trained model is considered an outlier to the *B* space. The *B* space was created automatically by the isolation forest algorithm using the same training data as for the neural network, more specifically, the outputs of the neural network.

5 Conclusions and Future Work

Based on the experimentation, a neural network with one hidden layer can fit correctly the MFCC processed from the vibration data of the motor to the MFCC processed from the vibration data of the end effector.

This algorithm is a good way of avoiding long mathematical models which can add errors, and which must be reworked each time a new component is added, or the kinematic chain is changed. The algorithm is suitable to any system that follows the specific pattern presented in Fig. 5.

As future work, it would be interesting to use the same data with a classifying neural network to also diagnose faults when enough data is available. Also, the regression neural network presented in this paper will be integrated in a larger monitoring system, being part of the system that detects the faults.

Acknowledgment. This work has been funded by the European Social Fund from the Sectoral Operational Programme Human Capital 2014–2020, through the Financial Agreement with the title "*Scholarships for entrepreneurial education among doctoral students and postdoctoral researchers (Be Antreprenor!)*", Contract no. 51680/09.07.2019 - SMIS code: 124539.

References

1. Costa, B.S.J., Angelov, P.P., Guedes, L.A.: Fully unsupervised fault detection and identification based on recursive density estimation and self-evolving cloud-based classifier. Neurocomputing **150**, 289–303 (2015)
2. Silvestri, G., Verona, F.B., Innocenti, M., Napolitano, M.: Fault detection using neural networks. In: Proceedings of IEEE International Conference on Neural Networks (1994)
3. Nelwamondo, F.V., Marwala, T. : Faults detection using Gaussian mixture models, mel-frequency cepstral coefficients and kurtosis. In: Conference Proceedings - IEEE International Conference on Systems, Man and Cybernetics (2007)
4. Singh, V., Meena, N.: Engine fault diagnosis using DTW, MFCC and FFT. In: Proceedings of the First International Conference on Intelligent Human Computer Interaction (2009)
5. Shaughnessy, O.: Speech Communications: Human And Machine, 2nd edn. Universities Press, Hyderabad (1999)
6. Kingma, D.P., Ba, J.L.: Adam: a method for stochastic optimization. In: 3rd International Conference on Learning Representations, ICLR 2015 - Conference Track Proceedings (2015)
7. Liu, F.T., Ting, K.M., Zhou, Z.H.: Isolation forest. In: Proceedings - IEEE International Conference on Data Mining, ICDM (2008)

Computerized Techniques for Analysis of Lower - Limb Prostheses

Oana Andreea Chiriac(✉) and Doina Bucur

University POLITEHNICA of Bucharest, 313, Spl. Independentei, Bucharest, Romania
oana.chiriac27@gmail.com

Abstract. The main cause of limb loss is inadequate circulation in the limb owing to arterial disease, with more than half of all the amputations occurring among people with diabetes mellitus. Amputation of a limb may also occur after a traumatic event or for the treatment of bone cancer. A prosthesis is defined as "a device attached to the stump of an amputated body part due to traumatic or congenital conditions…" [1]. Lower limb prosthetics are devices designed to replace the function or appearance of the missing lower limb as much as possible.

Over time, have been developed various procedures for evaluating lower limb prostheses. The criteria used to assess comfort can be of two types: subjective, based on the technique of questionnaires that depend on the patient's mood and objective measures, based on variations in measurable physical quantities recorded by instrumentation and mathematical formulas that quantify comfort.

The main objective of the submitted article is to provide a clear summary of computerized techniques for analysis of lower-limb prostheses.

Keywords: Lower-limb prostheses · Computerized techniques · Biomechanics

1 Introduction

Lower limb prosthetics are devices designed to replace the function or appearance of the missing lower limb as much as possible. The basic categories of lower limb prostheses are, by the amputation height, transtibial (TT) and transfemoral (TF) prostheses. The typical transtibial prosthesis consists of a prosthetic foot, tube adaptor, and transtibial socket; a transfemoral prosthesis consists of a prosthetic foot, tube adaptor, a prosthetic knee joint, and transfemoral socket [2].

Lower limb prostheses can be classified into two categories: exoskeletal (prosthesis with the ability to support peripheral weight, the use of which facilitates the transfer of weight of a patient on the ground along the circumference of the device) or currently the most commonly used endoskeletal - modular prosthesis with the central weight - the support capacity, the use of which facilitates the transfer of a patient's weight to the ground, a tubular structure in the centre of the prosthesis) [3].

G. I. Gheorghe (Ed.): ICOMECYME 2020, LNNS 143, pp. 123–129, 2020.
https://doi.org/10.1007/978-3-030-53973-3_13

2 Biomechanics of Physiological Gait

The biomechanics of human movement can be defined as the discipline that describes, analyzes, and evaluates human movement. The physical movements involved are of large diversity: the gait of people with disabilities, the lifting of weight by a worker or the performance of an athlete. The physical and biological principles that apply are the same in all cases, what changes is only the specifics of the movement and the level of detail required in terms of the performance of each movement [4].

Human gait is achieved using a technique called the double pendulum (Fig. 1). From a kinetic perspective, it is a translational body movement in which the lever-rotary movement of the lower limb segments is transferred to the pelvic joint movement. In the forward movement, the leg leaves the ground and passes forward from the hip. This curve is the first pendulum; the heel now touches the ground and rolls away towards the toe in a motion described as the inverted pendulum [5].

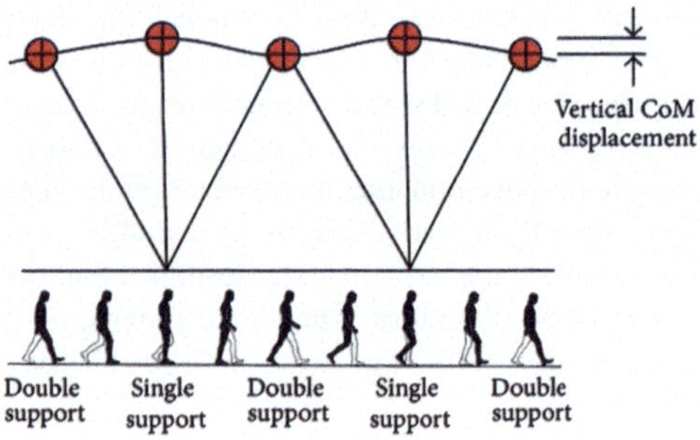

Fig. 1. Inverted pendulum model of gait, showing how the center of body mass (CoM) rises during the single support and falls during the double support [6].

A gait cycle is a period beginning with the initial contact between the heel and the ground of one leg up to the subsequent contact between the heel and the ground of the same leg (Fig. 2). A basic unit of the human gait is a step which is divided into 2 basic phases: the support phase and the swing phase [7].

In the support phase, a foot touches the ground and it takes approximately 60% of the overall cycle duration. The support phase can be double (in the beginning and in the end of the cycle), when the support is provided by both limbs, and a single, when only one limb touches the ground. In the double support, both limbs touch the ground.

The swing phase begins in the end of the support phase, immediately after the push-off, and takes 40% of the gait cycle. In the swing, knee flexors are initially activated, they adduct the lower leg onto the ground. The balance of the swinging limb is thus disturbed which causes the swing of the entire lower limb forward [7].

Walking is the fundamental and the most important phenomenon in space and time, reflecting locomotor characteristics of an individual. It is characteristic of orthogonal body control, concurrent bending of the body, head, and upper limbs, and the method of using lower limbs.

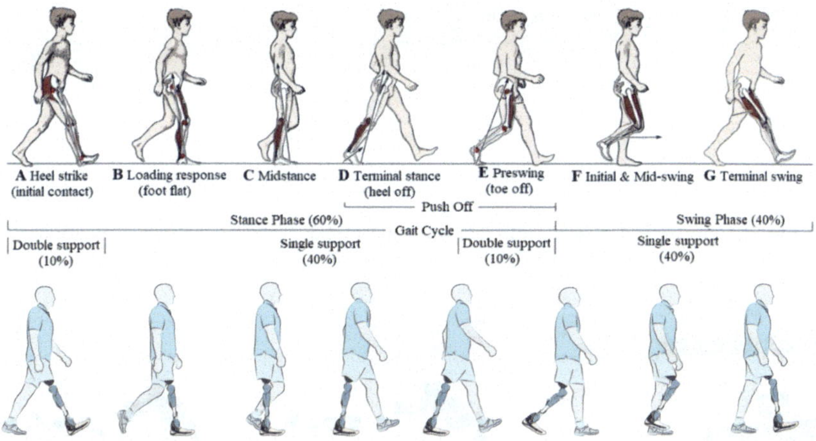

Fig. 2. Normal gait cycle vs. gait cycle with prostheses [7]

3 Equipment for Record-Keeping Motion and Measuring Plantar Pressure

The increased interest in the study of human movements has led to the establishment of recording techniques and equipment for these techniques. These allowed the analysis of movement in terms of quality (an aspect of movement) or quantitative (speeds, accelerations, forces) contributing significantly to the understanding of the fundamental mechanisms underlying human movement.

This chapter proposes an analysis of the distribution of plantar pressure with two types of equipment from different manufacturers. Novel - Pedar system and Tekscan – F-scan system are used. For each subsequent recording and analysis, a measurement procedure has been established both for the Pedar system and the Tekscan system, following the same steps for recording and analyzing the plantar pressure. Two men are evaluated, one with a TT prostheses (left) and the other with a TF prostheses (right) (Figs. 3, 4, 5, 6, 7, 8, and 9).

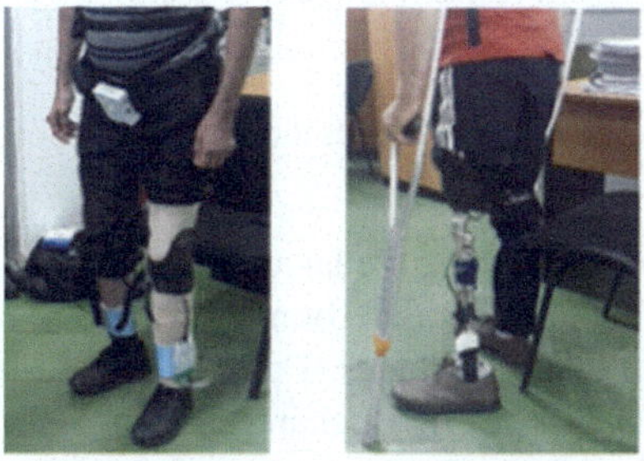

Fig. 3. TT prostheses and TF prostheses

pedar Files 6

Fi	Contact area (L) cm²	Contact area (R) cm²	Contact time (L) ms	Contact time (R) ms	Max force (L) N	Max force (R) N	Peak pressure (L) kPa	Peak pressure (R) kPa
C:\	133,30 ± 8,11	122,77 ± 33,73	1404,6 ± 1537,3	1576,0 ± 878,9	761,75 ± 18,74	751,83 ± 257,65	156,0 ± 12,3	332,2 ± 80,0
C:\	136,75 ± 7,94	138,99 ± 4,47	1110,5 ± 1031,3	1550,0 ± 536,5	810,09 ± 55,86	893,69 ± 37,49	161,3 ± 9,2	340,6 ± 48,1
C:\	156,50 ± 5,65	133,73 ± 33,09	866,3 ± 71,1	1304,7 ± 302,0	854,78 ± 38,75	867,43 ± 210,87	160,2 ± 5,0	327,5 ± 48,1
C:\	172,16 ± 5,43	142,21 ± 4,55	953,7 ± 173,3	1212,0 ± 628,1	949,37 ± 35,56	839,17 ± 35,79	187,9 ± 13,1	275,6 ± 62,6
C:\	156,51 ± 3,71	142,42 ± 4,52	941,1 ± 308,6	1060,0 ± 296,3	932,16 ± 61,40	851,13 ± 25,94	190,1 ± 11,5	295,8 ± 71,1
C:\	169,87 ± 7,92	145,09 ± 3,03	855,0 ± 84,7	1014,7 ± 63,9	1001,07 ± 41,36	851,09 ± 19,95	197,8 ± 10,5	343,2 ± 47,0

Fig. 4. Numerical results obtained with the Pedar system for the TT prosthesis

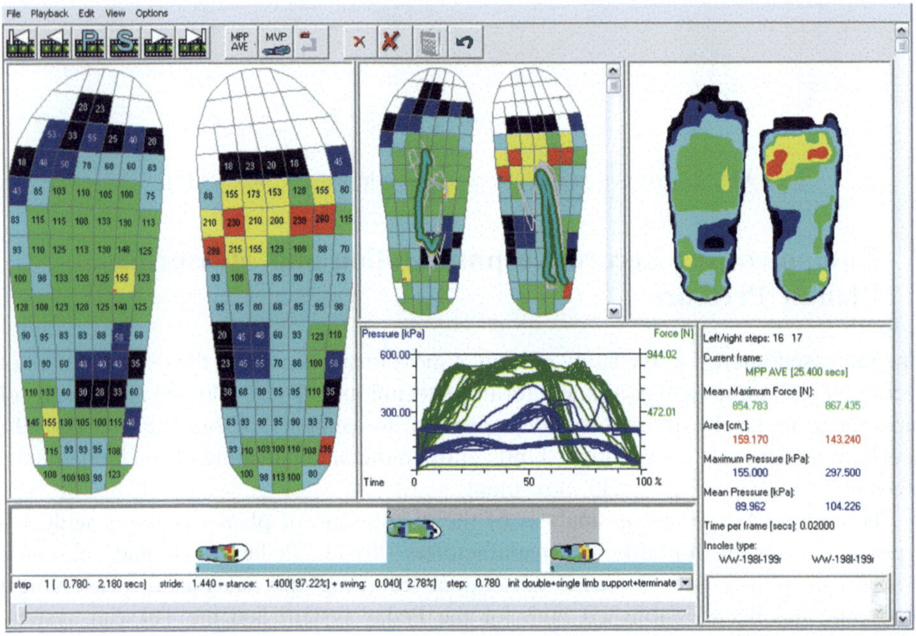

Fig. 5. Plantar pressure distribution and visualization of the walking line (TT prosthesis)

Fi	Contact area (L) cm²	Contact area (R) cm²	Contact time (L) ms	Contact time (R) ms	Max force (L) N	Max force (R) N	Peak pressure (L) kPa	Peak pressure (R) kPa
C:\	114,52 ± 51,47	45,35 ± 29,41	2370,0 ± 2077,1	1098,0 ± 780,6	287,81 ± 164,54	173,35 ± 98,29	129,4 ± 61,3	177,0 ± 40,6
C:\	100,74 ± 47,71	58,79 ± 30,29	1801,8 ± 1293,6	1118,0 ± 539,3	245,62 ± 144,54	268,35 ± 119,14	104,3 ± 52,2	228,0 ± 66,8

Fig. 6. Numerical results obtained with the Pedar system for the TF prostheses

The measurements, both with the Pedar equipment (Novel) and with the F-scan equipment (Tekscan) were made in the Biomechanics Laboratory of the Faculty of Mechanical Engineering and Mechatronics (Department of Mechatronics and Precision Mechanics), University Politehnica of Bucharest.

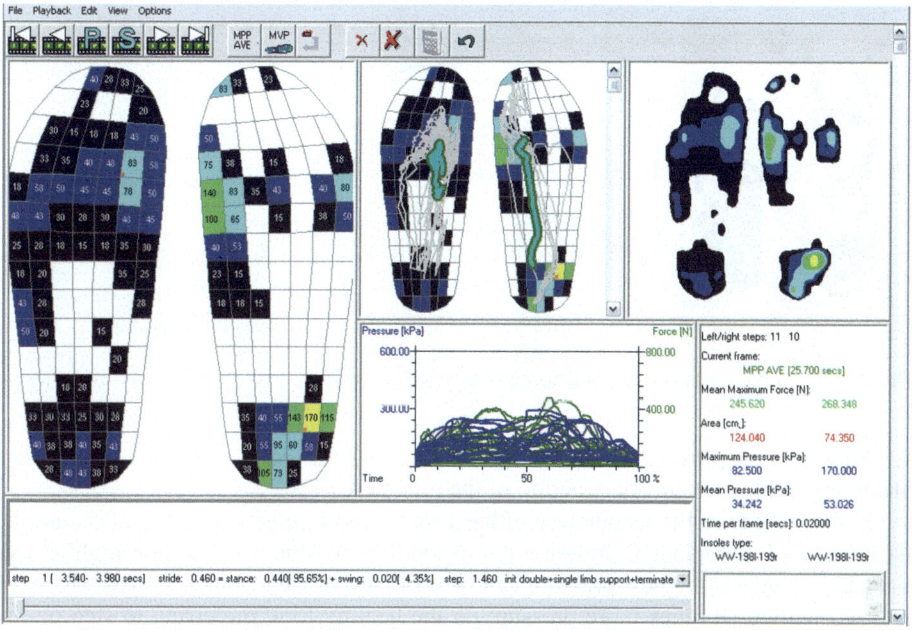

Fig. 7. Plantar pressure distribution and visualization of the walking line (TF prostheses)

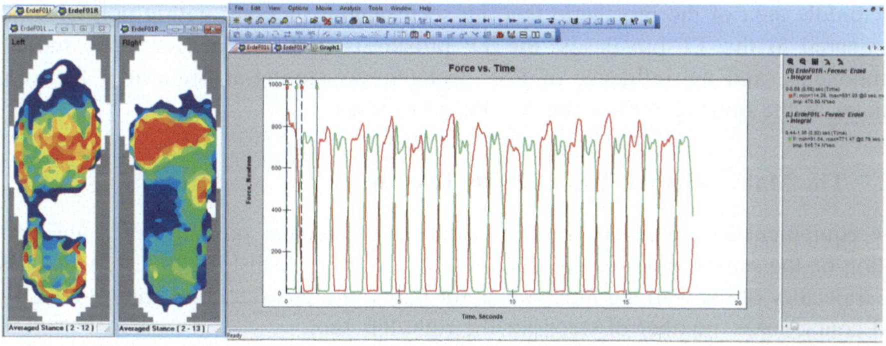

Fig. 8. 2D representation (left) and force overtime, on walking cycles (right) (TT prosthesis)

3.1 The Main Achieved Results with Pedar

To record gait and the value of the plantar pressure, the subject is asked to walk on a straight path, performing a series of gait cycles. The results can be displayed in different forms: plantar pressure distribution on the insole; 3D pressure distribution; 2D pressure distribution; step analysis; values for maximum pressure, maximum force, time and contact area, number of steps for the right/left foot.

It should also be mentioned that the person has an amputation of the toes (toes 1 and 2) on the right foot. From the first records, can be observed a uniform plantar pressure

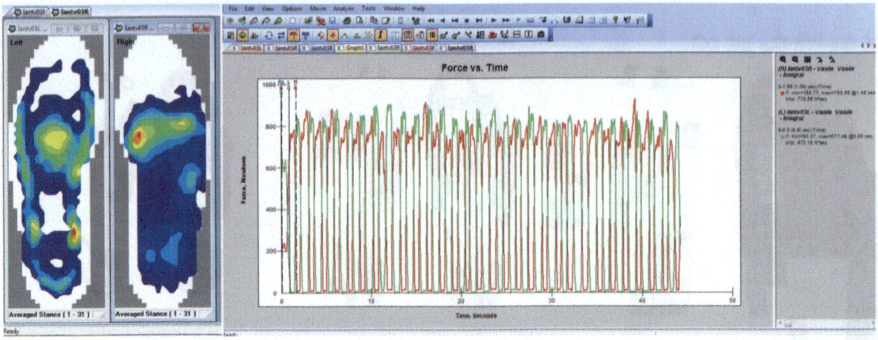

Fig. 9. 2D representation (left) and force overtime, on walking cycles (right) (TF prosthesis)

distribution for the prosthetic foot (left foot). The prosthetic foot does not run on the entire sole, being limited by the construction of the prostheses.

On the right foot, the amputation of toe 1 and 2 can be easily observed, because the distribution of the plantar pressure corresponding to this area is not highlighted. Instead, the range of highest pressure can be observed in the metatarsal areas, since the analyzed individual relies more pressure on the healthy foot, preferring to step on the tip of the foot to maintain a balance in dynamic and static conditions.

The recordings made with the TF prostheses show a uniform distribution of the plantar pressure, being able to state that the prosthetic foot runs like a healthy foot. On the middle area of the prosthetic feet are seen plantar pressure within normal limits. Compared to the TT prosthesis, the TF prosthesis uses high-performance modular components, and the influence of ESR feet (energy-storing-and-returning feet) contributes to the optimal performance of the prostheses.

3.2 The Main Achieved Results with Tekscan

The equipment can represent the next information: plantar pressure; pressure distribution on the sensor; 3D pressure distribution; 2D pressure distribution; step analysis - the trajectory of the walking line; values for maximum pressure, maximum force, time and contact area, number of steps for the right/left foot.

The force value for the right leg (non-prosthetic leg) is higher compared to the left leg (prosthetic leg). Consequently, it can be confirmed that the patient rests on the right foot, although this foot has amputation at the toes 1 and 2. This is thanks to the modular prosthesis with standard components, which does not give the non-prosthetic foot stability.

The trajectory of the running line is strongly influenced by the choice of modular components, which allow a series of improvements so that a proper adjustment can be achieved. The walking line is also influenced by the choice of the prosthetic foot, which in this case gives stability and propulsion to the analyzed subject. Even if no conclusions can be made regarding the actual values of the plantar pressures, the distribution of the plantar pressure can be studied from a qualitative point of view.

As a result, it can be easily observed from the images obtained that the analyzed subject walks with the entire prosthetic sole (prosthetic foot) having a uniform gait line similar to the anatomic gait line.

4 Conclusions

Proper fitting of the prosthesis is influenced by normal walking stereotyping of the patient, the role of the prosthetic foot and the pressure on the stump has a direct effect on the comfort and energy consumption in the use of the prostheses. The wrong prosthetic fitting can cause pain to users during the execution of daily activities. The manifestation of pain can correspond to the lateral asymmetry of the body caused by the incorrect length of the prosthesis or incorrectly selected components. Wrong construction of the prosthesis can lead to an imbalance of forces, overload muscle groups, risk of tripping and also to damage of soft tissues integration on the stump.

This study aimed to investigate the optimization of TT prostheses and TF prostheses, from a constructive point of view, using plantar pressure distribution analysis software.

The impact of the distinct types of a prosthesis (TT prosthesis and TF prosthesis) was studied, as well as the impact of the distribution of plantar pressure on the prosthetic foot and the healthy foot. At the same time, were studied parameters such as the walking line, the maximum and minimum value of plantar pressure and force, as well as the surface of the contact area.

References

1. John, C.: Three Years of Work for Handicapped Men - A Report of the Activities of the Institute for Crippled and Disabled Men, pp. 19–24. The Institute, New York (1999)
2. Bowker, J.H., Michael, J.W.: Atlas of Limb Prosthetics: Surgical, Prosthetic and Rehabilitation Principles, 2nd. edn. American Academy of Orthopeadic Surgeons, p. 930 (1992). ISBN 0-8016-0209-2
3. Lee, R.Y., Turner-Smith, A.: The influence of the lenght of lower-limb prosthesis on spinal kinematics. Arch. Phys. Med. Rehabil. **84**, 57–62 (2003)
4. Biomechanics course notes: Biomechanics as an interdisciplinary, First Year Master, Faculty of Medical Engineering, University Politehnica of Bucharest, pp. 1–48 (2018)
5. Mak, A.F.T., Zhang, M., Boone, A.C.P.: State of the art research in lower limb prosthetic biomechanics socket interface. J. Rehabil. Res. Dev. **38**(2), 161–174 (2001)
6. Lobet, S., Detrembleur, C., Massaad, F., et al.: Review article. Three-dimensional gait analysis can shed new light on walking in patients with haemophilia. Sci. World J. **2013**, 1–7 (2013). Article ID 284358, Research Gate
7. Rajtukova, V., Michalíková, M., Balogova, A., Zivcak, J.: Biomechanics of lower limb prostheses. Procedia Eng. **96**, 382–392 (2014). Technical University of Kosice, Faculty of Mechanical Engineering, Department of Biomedical Engineering and Measurement. ResearchGate

From Conventional Prosthetic Feet to Bionic Feet. A Review

Oana Andreea Chiriac(✉) and Doina Bucur

University POLITEHNICA of Bucharest,
313, Spl. Independentei, Bucharest, Romania
oana.chiriac27@gmail.com

Abstract. A prosthesis is defined as "a device attached to the stump of an amputated body part due to traumatic or congenital conditions…" [1]. Prostheses have evolved in recent centuries, at first, they were made of wood but specialists in the field have conducted research to develop new materials and technologies, such as carbon fiber foot or bionic ankle joint. Generally, prosthetic feet can be divided into three categories. According to the schedule presented into the article, they are regular feet (CF), energy storage and return (ESR) feet, and the so-called "bionic" feet.

Each prosthesis is designed and assembled according to the person's physical appearance, functional needs, and accessibility. To this day, the design of the leg prosthesis has improved, but it cannot actively adapt to different walking speeds in a way comparable to biological limbs. Since the 1990s, more and more attention has been paid to the incorporation of active components in prosthetic limbs because passive devices cannot provide enough power for the ankle while walking. Today, most of these bionic devices are still at the research level, but they can be expected to be on the market soon. In this paper, the evolution of prosthetic feet over the last two decades is reflected, from early prosthetic feet to recent bionic feet.

Keywords: Foot prostheses · Energy-storing-and-returning · Bionic feet

1 Introduction

The design and manufacture of lower limb prostheses is closely related to the medical options regarding the level of amputation, but also to the range of existing prosthetic materials and components. In the past, due to absence of accessible biocompatible materials and complex surgical techniques, the only solution was to remove the affected tissue, amputate and replace, where appropriate, with an external prosthesis. Currently, the field of prosthesis has evolved, developing new materials and technologies for making high-performance lower limb prostheses (Figs. 1 and 2).

Approximately 80% of all lower limb amputations are caused by peripheral vascular diseases (including diabetes), which is increasingly affecting the population (55–75 years), socially active both in the developed countries and the ones least developed.

The principle of prosthesis is the use of artificial parts (prostheses) to improve the vital function and lifestyle of people with motor disabilities and more. It is well known

G. I. Gheorghe (Ed.): ICOMECYME 2020, LNNS 143, pp. 130–138, 2020.
https://doi.org/10.1007/978-3-030-53973-3_14

that the loss of a unilateral or bilateral segment, as well as various injuries that affect a physiological function or create a vicious attitude of the musculoskeletal system, give rise to a mental trauma with profound reactions on the individual. Completing an amputated limb is one of the most important goals in the field of prosthetics and is also an independent part of biomechanics and therapy. Prostheses are complex, can vary depending on their applicability and can be classified according to vastly different criteria [2, 3].

A critical objective in the field of prosthetic leg design is to advance an ankle – foot prosthesis capable of emulating the dynamics of the biological ankle [4]. On the one hand the use of carbon fiber ankle minimizes the weight of the prosthetic limb and is particularly important for the amputee. The comfort and functionality of the prosthetic limb are highly dependent on its weight. This includes reducing the weight of the socket which attaches to the residual limb, and to the various connectors and struts comprising the total prosthetic limb. The most critical areas where weight should be reduced are those on the distal portion of the prosthetic limb, the foot itself. On the other hand, the aim of powered ankle is to improve the mobility of individuals with below knee amputations by closely imitating the biomechanical function of the missing biological limb.

This article will present what is currently known about carbon fiber prostheses with built-in ankle joint as well as new research on the powered ankle and the evolution from early prosthetic feet to recent bionic feet is presented bearing in mind the importance of human ankle biomechanics.

2 The Development of Carbon Fiber Ankle – Foot Prosthesis

Numerous prosthetic feet are currently on the market for individuals with a transtibial amputation, each device aimed at raising the 3C-level (control, comfort, and cosmetics) with slightly distinctive characteristics. In general, prosthetic feet can be classified into three categories. Versluys et al. (2009) classify the recent timeline of prosthetic feet into three categories: conventional feet (CF), energy-storing-and-returning (ESR) feet and the recent so-called 'bionic' feet [5].

2.1 Conventional Feet (CF)

SACH Foot

It was not until the 1980s that the design of prosthetic feet began to restore basic walking and allow amputees to complete basic tasks. The prostheses evolved to the conventional feet (CF), which were still considerably basic but allowed future prostheses to focus on weight and functionality [8].

Early designs for prosthetic feet were often a solid piece of wood. A similar design, the SACH (solid-ankle-cushioned-heel) is still in use because of its sturdy function, especially useful for individuals with lower activity levels. A SACH foot typically has a rigid inner structure (wood or plastic) surrounded by a compressible foam cosmetic shell [8].

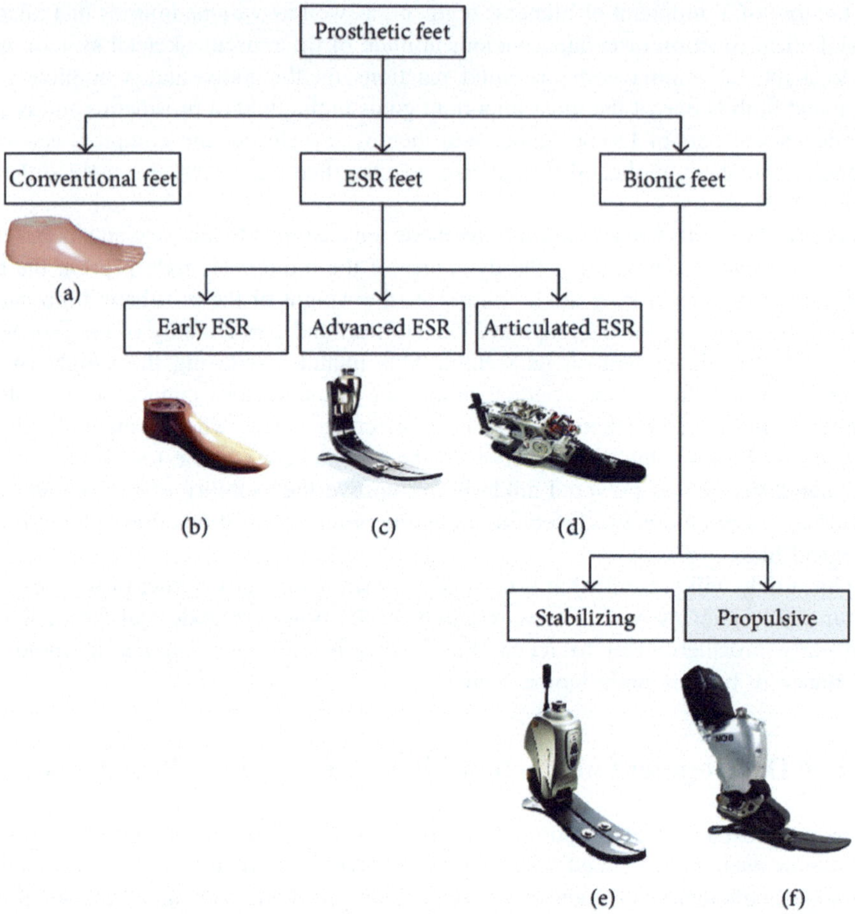

Fig. 1. Categorization of today's prosthetic feet with as example (a) the SACH foot, (b) the SAFE foot, (c) Ossur's Flex-Foot, (d) the CESR foot [6], (e) Ossur's Proprio Foot, and (f) iWalk's Powerfoot BiOM [7].

Different models are still available. The heel cushion compresses at heel contact to mimic plantarflexion and establish foot flat. Flexible toe portion replaces some of the dorsiflexion in mid to late stance.

Driven by the amputees' desire to walk more naturally, lowering metabolic costs and even the desire to exercise in certain circumstances, the prosthetic leg has improved significantly in recent decades, leading to the development of energy stock and recovery (ESR) and recent developments.

2.2 Energy-Storing-and-Returning (ESR)

The desire of amputees to participate in sports led to the development of early ESR feet, which store energy by loading a spring with the body weight. The energy lost in the form of friction in the system is high and dissipated in the form of heat and sound [5, 9].

Early ESR

Conventional feet like the SACH foot [8] can restore basic walking ability by improving stability and providing a roll-over to make walking more comfortable. The early ESR-feet like the Seattle foot and many more (the Dynamic Plus foot, the C-Walk, and the Carbon Copy foo) [10, 11] are different from the conventional feet because they store energy early in the gait cycle and release this energy at push-off when it is needed to move the body forward.

The letters S.A.F.E. are the acronym for Stationary Attachment Flexible Endoskeleton or, in simpler language, a prosthetic foot bolted to the shin with a flexible keel. In order to maintain consistency with endoskeletal systems, the foot's bolt block and flexible keel are wrapped in a soft foam cover [12].

The foot must not only support weight or compressive force, but also must have a dynamic motion force that generates torque and shear force. However, apart from the lack of an ankle, it may describe the human foot in a simple way.

Advanced ESR

Advanced ESR feet have better properties than early ESR feet [5, 9]. Many active individuals usually wear energy storage and return feet (ESR), also known as "dynamic elastic response" (DER) [13]. This foot enhances the comfort of the disabled; however, the ability of the ESR foot to mimic the human ankle-foot complex is limited. The amputee's gait speed is still slower than normal walking [14], and many studies have shown that ESR feet have not resolved the increased metabolic energy consumption that leads to early fatigue [15, 16]. In addition, natural walking patterns are still severely disturbed, leading to long-term medical conditions [17, 18].

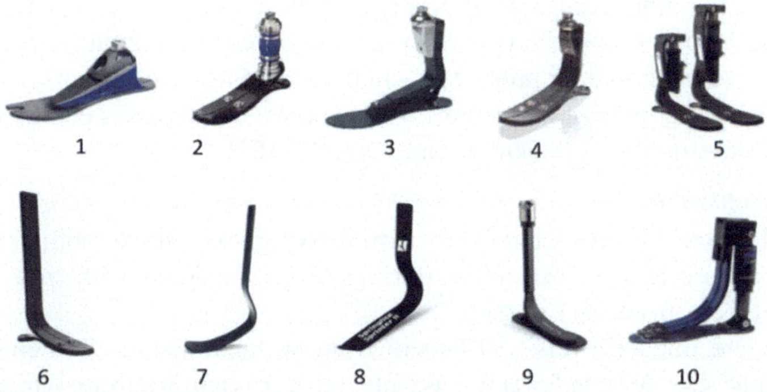

Fig. 2. Advanced ESR feet. (1) Flex-Foot Axia. (2) LP-Ceterus. (3) Talux Foot. (4) VariFlex. (5) Re-Flex VSP. (6) Flex-Foot ModularII. (7) Flex-Sprint Cheetah. (8) Sprinter. (9) Advantage DP/Springlite Foot. (10) Pathfinder [5].

Flex-Foot Axia

Flex-Foot Axia is a multi-axis foot that provides improved terrain consistency and guided rollover response. Designed to replicate the movements of the anatomical foot, the guided rollover function increases the lateral stability of the foot, enabling more balanced movement during standing and increasing user comfort [19].

It provides a multi-axis range of motion, with a moderate energy response and a low height, so it is an ideal choice for people with medium to long residual limbs who need a comfortable gait and consistent gait [19].

LP-Ceterus

Recommended for amputees with long residual limbs, the building height of this excellent energy storage foot is 70–82 mm lower than the standard Ceterus model. LP Ceterus has the same progressive rotation and adjustable vertical shock absorption, including a new foot module that provides greater flexibility and energy return than any previous low-profile design [20].

Talux Foot

From heel strike to toe-off, no other prosthetic foot simulates a smoother or more graceful walking movement [21].

Talux mimics many anatomical features of the human foot and was specifically developed to provide users with low to moderate activity with a smooth and natural walking movement in various terrains. Talux feet are also equipped with sandal toes for quick footwear selection [21].

VariFlex

Vari-Flex is lightweight and easy to assemble, so whether it is a prosthesis or a user, it is a natural choice. It is available in both male and female versions. It can be used with conventional 30 mm graphite towers and male pyramids, making it simple to add to endoskeletal pylon components [22].

Re-Flex VSP

The composite spring in front provides the best shock absorption and reduces the impact on the body. Increase control and comfort during walking and other activities.

It uses a carbon fibre compression spring and telescopic tubes that provide a vertical compression of up to one inch. This cushions the impact to amputees' residual limbs, allowing them to land on their prosthesis when descending stairs or curbs, for example, without experiencing pain or discomfort. It also stores and releases energy to allow the user to walk comfortably, efficiently, and naturally [23].

Flex-Foot Modular II

Flex-Foot Modular II is characterized by extremely lightweight, durability, high energy storage and release feet. 100% carbon fibre provides amputees with smooth and continuous movement from heel to toe.

All ages and impact levels will benefit from an unparalleled 95% energy storage and return. It is very durable for active people, but very light for those who are not very active. The people who can benefit from this foot are unlimited. This foot can be customized for almost all weight restrictions [24].

Flex-Sprint Cheetah
Flex-Foot Cheetah is an artificial foot replacement developed by biomedical engineer Van Phillips. He lost his leg below the knee when he was 21. The deficiency of the existing prosthesis led him to invent this new prosthesis.

A customized high-performance carbon-fibre foot, mainly used for sports activities such as sprinting. With a unique J curve design, it can maximize the energy feedback [25].

Sprinter
The Sprinter foot can be combined with other components that have been designed and tested for performance and can be used for running or similar sports activities. The feet provide very high energy feedback, and there are six versions of stiffness that can adapt to various weights. It is characterized by lightweight. The spring profile ensures high driving force and low resistance [26].

Springlite Foot/Advantage DP
Springlite Foot gives a better dynamic response for recreational athletes thanks to the pylon foot. The advantage of this kind of foot is the DP2 hanger, its outstanding dynamic response and good shock absorption performance make it unique.

The pylon foot are approved to weigh up to 150 kg, and are suitable for prosthetic wearers who undergo amputation of the femur or tibia and dislocation of the hip or knee [27].

Pathfinder Foot
Pathfinder foot made by Willow Wood, Ohio are dynamic feet with high stability and are suitable for highly active amputees. Pathfinder has unique toe spring, pedal structure and adjustable pneumatic heel spring, its triangular structure can provide high energy return, rotation, reverse and eversion [28].

Articulated ESR
Articulated ESR feet usually use electronic equipment and small servo motors to engage or disengage the locking mechanism. Although these devices are highly precise electromechanical systems, they do not use external mechanical power to provide active stability or sagging characteristics to the wearer, so they are outdated in the use of foot propulsion bionic technology [7].

In 2010, researchers at the Vrije Universiteit Brussel, Belgium, have developed the Ankle Mimicking Prosthetic Foot (AMP-Foot 1.0) [29], an articulated ESR-type foot. Compared to traditional ESR feet used as torsion springs, articulated ESR is one of the first systems to use a locking mechanism to store and release the extracted energy during the dorsiflexion (DF) phase of the posture. The so-called Controlled Energy Storing and Returning Foot (CESR Foot) was developed to enhance the push-off properties of passive prostheses. CESR feet do not store energy during posture, but use the weight of the body to accumulate energy during initial contact and release energy when needed [30].

All the prostheses mentioned use only the energy generated by the amputee of its own, to imitate a healthy ankle's behavior.

2.3 Bionic Feet

Bionic foot is defined as a mechanical device with one or more active ingredients used to stabilize the foot or provide active sagging characteristics (advancing the bionic foot), that is, worn by a person with an TT amputation [7]. Most commercial tibial prostheses today use actuation to help stabilize the ankle complex.

Examples are the Motion and Raize Foot (Fillauer), the Elan foot (Endolite), and the Proprio Foot (Ossur). This type of prosthesis uses hydraulic or electric actuation to provide natural ankle kinematics and intelligent terrain adaptability, but the power provided does not even exceed the power stored during gait.

3 Conclusions

In the previous sections, an update of prosthetic feet technology is provided. The evolution from early conventional feet to recent bionic feet is presented bearing in mind the importance of human ankle biomechanics.

In general, prosthetic feet can be divided into three categories: conventional feet (CF), energy storage and return energy (ESR) feet, and bionic feet. Compared with the CF foot, the ESR foot can store energy in the elastic element and return to its main part to help propel the propulsion force. Amputees still prefer advanced ESR feet other than CP. Of course, this is purely an issue of comfort as it is clearly seen that, despite advanced engineering and manufacturing technology, no prosthetic foot provides energy return that results in statistically significant decreasing metabolic and improved gait.

Due to technological advances, many artificial feet with enhanced mechanical properties have appeared in the past two decades. In addition to aesthetic purposes, in the design of prosthetics, the biomechanical properties of the ankle-foot structure have received more attention.

Nowadays, it is strongly believed that new prosthetic feet concepts should be configured with active components to completely describe human ankle function.

References

1. John, C.: Three Years of Work for Handicapped Men - A Report of the Activities of the Institute for Crippled and Disabled Men, pp. 19–24. The Institute, New York (1999)
2. Marks, L.J., Michael, J.W.: Clinical review artificial limbs. BMJ **323**, 732–735 (2001)
3. Stryła, W., Pogorzała, A.M., Kasior, I., Nowakowski, A.: Limb amputations from the ancient times to the present. Pol. Orthop. Traumatol. **78**, 155–166 (2013)
4. Herr, H.M., Grabowski, A.M.: Bionic ankle – foot prosthesis normalizes walking gait for persons with leg amputation. Proc. R. Soc. **279**, 457–464 (2011)
5. Versluys, R., Beyl, P., Van Damme, M., Van, D.A., Ham, R.: Prosthetic feet: state-of-the-art review and the importance of mimicking human ankle-foot biomechanics. Disabil. Rehabil. Assist. Technol. **4**(2), 65–75 (2009)

6. Collins, S.H., Kuo, A.D.: Controlled energy storage and return prosthesis reduces metabolic cost of walking. In: Proceedings of the ISB 20th Congress—ASB 29th Annual Meeting, p. 1 (2005)
7. Cherelle, P., Mathijssen, G.: Review article. Advances in propulsive bionic feet and their actuation principles. Adv. Mech. Eng. **2014**, 1–22 (2014)
8. Ottobock Homepage, Sach foot. https://www.ottobock.co.uk/prosthetics/info_for_new_amputees/prosthetic-technology-explained/about_feet/index.html. Accessed 12 May 2020
9. Brackx, B., VanDamme, M., Matthys, A., et al.: Passive ankle-foot prosthesis prototype. Int. J. Adv. Robot. Syst. **10**, 1–9 (2013)
10. Arya, A.P., Lees, A., Nerula, H.C., Klenerman, L.: A biomechanical comparison of the SACH, Seattle and Jaipur feet using ground reaction forces. Prosthet. Orthot. Int. **19**(1), 37–45 (1995)
11. Hafner, B.J., Sanders, J.E., et al.: Transtibial energy-storageand- return prosthetic devices: a review of energy concepts and a proposed nomenclature. J. Rehabil. Res. Develop. **39**, 1–11 (2002)
12. Campbell, J.W., Childs, C.W.: The S.A.F.E. foot. Orthot. Prosthet. **34**(3), 3–16 (1980). Digital Resource Foundation for the Orthotics & Prosthetics Community
13. Torburn, L., Perry, J., Ayyappa, E., Shanfield, L.S.: Below-knee amputee gait with dynamic elastic response prosthetic feet: a pilot study. J. Rehabil. Res. **27**(4), 369 (1990)
14. Colborne, G.R., Naumann, S., Longmuir, P.E., Berbrayer, D.: Analysis of mechanical and metabolic factors in the gait of congenital below knee amputees: a comparison of the SACH and Seattle. Am. J. Phys. Med. **71**(5), 272 (1992)
15. Torburn, L., Powers, C.M., Guiterrez, R., Perry, J.: Energy expenditure during ambulation in dysvascular and traumatic below-knee amputees: a comparison of five prosthetic feet. J. Rehabil. Res. Develop. **32**(2), 111–119 (1995)
16. Perry, J., Boyd, L.A., Sreesha, S.S., Mulroy, S.J.: Prosthetic weight acceptance mechanics in transtibial amputees wearing the Single Axis, Seattle Lite, and Flex Foot. IEEE Trans. Rehabil. Eng. **5**(4), 283–289 (1997)
17. Bateni, H., Olney, S.J.: Kinematic and kinetic variations of below-knee amputee gait. J. Prosthet. Orthot. **14**(1), 2–10 (2002)
18. Burke, M.J., Roman, V., Wright, V.: Bone and joint changes in lower limb amputees. Ann. Rheum. Dis. **37**, 252–254 (1978)
19. Ossur Homepage, Flex-Foot Axia. https://assets.ossur.com/library/19193/FLEX-FOOT. Accessed 13 May 2020
20. Ossur Homepage, LP Ceterus. https://assets.ossur.com/library/24638. Accessed 13 May 2020
21. Ossur Homepage, Talux Foot. https://www.ossur.com/en-gb/prosthetics/feet/talux. Accessed 13 May 2020
22. Ossur Homepage, Vari-Flex. https://www.ossur.com/en-gb/prosthetics/feet/vari-flex. Accessed 13 May 2020
23. Ossur Homepage, Re-Flex VSP. https://www.ossur.com/en-gb/prosthetics/feet/re-flex-shock. Accessed 14 May 2020
24. Ossur Homepage, Flex-Foot Modular II. https://assets.ossur.com/library/30499/Vari-Flex%20Modular%20Catalog%20page.pdf. Accessed 14 May 2020
25. Ossur Homepage Flex-Sprint Cheetah. https://assets.ossur.com/library/19207/Flex-foot%20cheetah®.pdf. Accessed 14 May 2020
26. Ottobock Homepage, Sprinter Foot. https://shop.ottobock.us/Prosthetics/Lower-Limb-Prosthetics/Fitness-Prosthetics/Springlite-Sprinter/p/1E90. Accessed 14 May 2020
27. Ottobock Homepage, Springlite Foot/Advantage DP. https://www.ottobock.com.tr/en/prosthetics/products-from-a-to-z/advantage-dp2-1e50-1e52/. Accessed 14 May 2020

28. Shankman, G., Manske, R.: Fundamental Orthopedic Management for the Physical Therapist Assistant, 3rd edn. Elsevier MOSBY, Maryland Heights (2011)
29. Brackx, B., Damme, M.V., Matthys, A., Vanderborght, B., et al.: Passive ankle-foot prosthesis prototype with extended push-off. Int. J. Adv. Robot. (in review) (2011)
30. Segal, A., Zelik, K.E., Klute, G.K., et al.: The effects of a controlled energy storage and return prototype prosthetic foot on transtibial amputee ambulation. Hum. Mov. Sci., 1–30 (2010)

Device for Injection Molding Realized by Additive Technologies

Elena Dinu, Daniel Besnea(✉), An Sebastian Ping, Alina Spanu, Edgar Moraru, and Iolanda Panait

University Politehnica of Bucharest, Splaiul Independentei no. 313, District 6, Bucharest, Romania
elena_dinu@yahoo.com, d_bes@yahoo.com, seby471@gmail.com, spanu_alina@yahoo.com, eddy_milan91@yahoo.com, iolanda_pan@yahoo.com

Abstract. The paper presents a way to obtain an injection device as well as making a mold using additive technologies and some rubber parts based on these molds. The design and the main parts of the injection device designed and made will be presented. This system can bring considerable advantages in terms of reducing the manufacturing costs of single series parts or small production without sacrificing the quality of the parts produced.

Keywords: Injection device · Mold · Additive technologies · Rapid prototyping

1 Introduction

Today it can be definitely said that 3D printing and 3D scanning are actively entering our lives and will soon become truly indispensable tools in many fields of activity. And if now only the specialists can evaluate the scale of development of this segment of the high-tech industry, then in the near future the situation will change. The 3D printing technique will be used much more widely and will penetrate our daily lives. Additive technology embodies the next industrial revolution leading to resource-efficient environmental production. These technologies can become even more widespread in injection molding processes.

Recently, additive technologies have been actively introduced in the world to overcome technological limitations and accelerate the design and production timelines. The creation of foundry molds by the methods of layer-by-layer synthesis allows to circumvent the technological limitations of traditional methods and shorten the production chain by abandoning long operations, which leads to a reduction in production time and an order of magnitude lower cost.

Molds made by RP usually have two standard configurations:

Mold inserts in aluminum frames (Fig. 1): this is the most common die configuration and generally produces much more precise parts. The mold is made by RP after which it is embedded in a rigid aluminum frame that ensures the support of the mold against the injection pressure and against the high temperature of the injection nozzle. This aluminum frame also prevents the mold from deforming after multiple uses.

G. I. Gheorghe (Ed.): ICOMECYME 2020, LNNS 143, pp. 139–148, 2020.
https://doi.org/10.1007/978-3-030-53973-3_15

Simple molds: in this configuration the mold is made completely by Rapid Prototyping and a rigid aluminum frame is no longer used. In this way, complex cooling channels can be integrated in the mold, but the molds made in this way require the use of more basic material, which increases the cost and time of making the mold and are also more prone to deformation after repeated use [1–4, 7].

Fig. 1. Mold embedded in a rigid aluminum frame [7]

2 Device Design

For the practical realization was designed the model that was made by Rapid Prototyping, a device for injection and support of molds made by standardized Rapid Tooling technologies to a size of 100 mm × 100 mm with a depth of half a mold of 20 mm, which allows the support of the mold in an aluminum frame, which also allows the opening of the molds using a screw with trapezoidal thread, which moves the moving half of the device on four columns of support, sliding on four bushes made of brass. Also, the moving part is guided by the feather made in the motherboard, this being rectified together with the supported face of the moving part to reduce friction between the two parts. The injection cylinder consists of a copper cylinder that can be positioned towards the injection mouth of the mold, it has a piston mechanism for pushing the molten plastic through the nozzle made of brass. The cylinder is heated by using 40 W resistive cartridges, connected to a RAMPS board, the control program used being a modified version of the Marlin open source firmware, modified to calibrate the calculation formulas for heating the copper cylinder, it being more difficult to heat compared to a normal 3D printer effector [6, 7] (Figs. 3 and 4).

The device consists of three main subassemblies: the injection piston holder, the mold closure system (Fig. 2) and the motherboard on which they are mounted. The entire assembly is mounted on this board by means of standard screws. It will support both the other two subassemblies and guide the movable plate of the mold during its closing and opening.

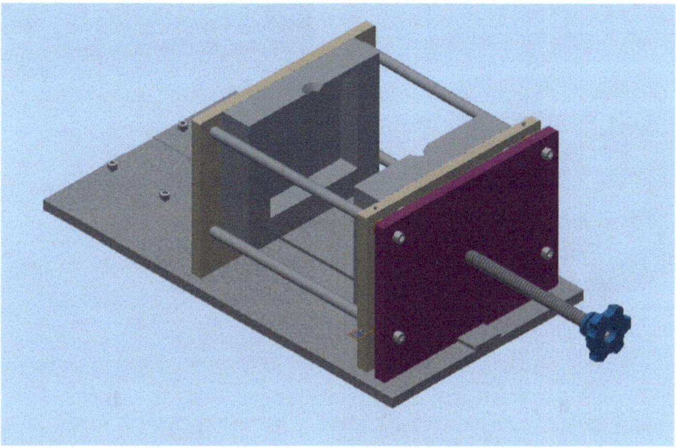

Fig. 2. Mold closing system

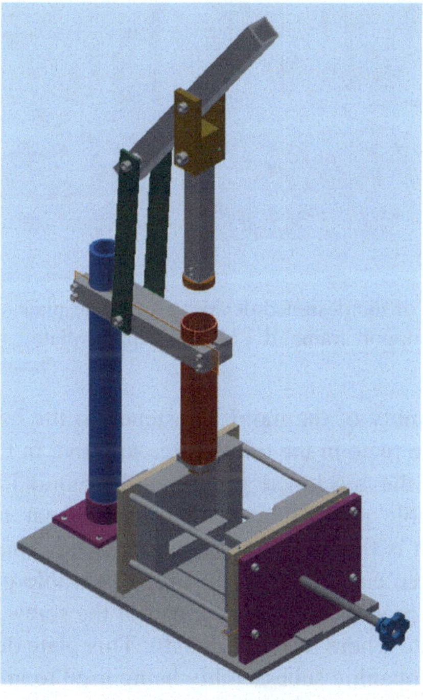

Fig. 3. Injection molding device made with RT (Rapid Tooling) technologies

This is due to the channel processed in the movable plate of the mold, which will be guided by the groove coming out of the center of the plate. In order to be as smooth as possible, both the channel in the movable plate and the groove in the base plate will be ground for the highest possible accuracy.

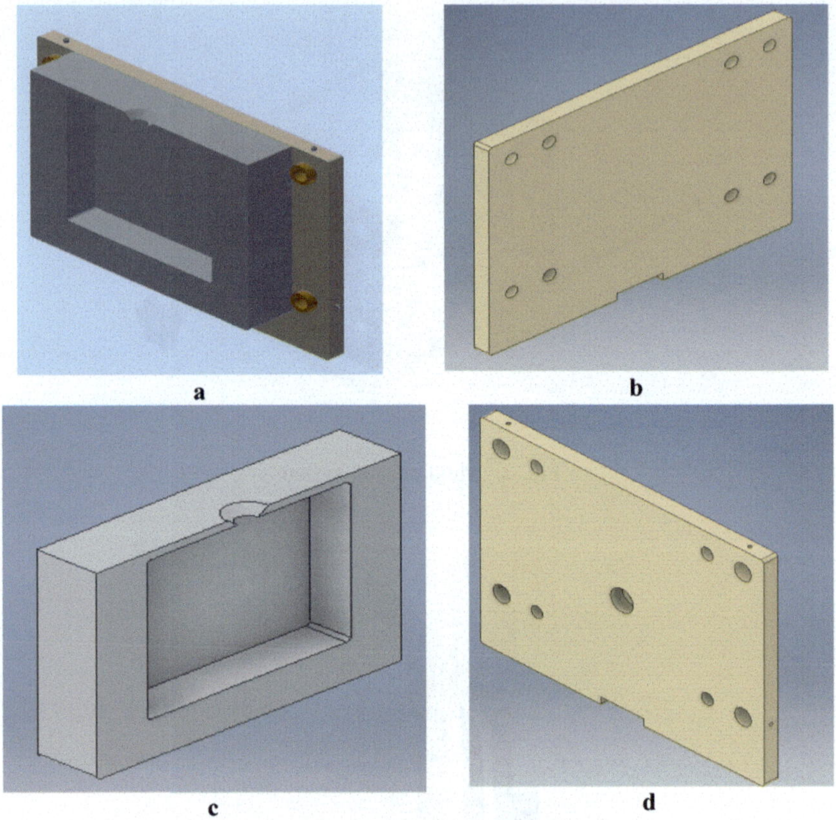

Fig. 4. Some components of the designed device: a - Mobile plate with mold holder mounted; b - Fixed plate; c - Mold support frame; d - Mobile support plate

The closing subassembly of the mold is fastened to the base plate in standardized screws, with the movable plate in the middle free to move in the longitudinal direction of the plate guided by the machined groove and ground in the middle of it. The movement of the movable plate is actuated by the linear movement made by the trapezoidal screw, which is threaded into the fixing plate. The opening of the mold is made by coupling the free end of the screw to the movable plate through a clamping ring placed in a channel processed in the free end of the screw, which has a processing made in the movable plate where it can be caught. This plate determines the position of the stroke end of the die opening subassembly being used to guide the linear movement of the trapezoidal screw, as well as to support the four guide columns on which the moving part of the device is guided. The columns can be made of standardized guide bar, because they are heat treated the outer surface for better mechanical resistance to frictional forces that may occur in these types of guides.

The movable plate is for supporting the support frame of the mold made by rapid prototyping, it has holes in it both for fixing the frame with screws, and for the brass bushes and their fixing screws. The face on which the Rapid Prototyping mold frame is mounted will need to be ground to ensure a perfect frame placement, which will reduce the possibility of problems aligning the mold halves. Also, the channel and the back face of the plate will have to be rectified, because these surfaces are used to guide the mobile assembly.

The fixed plate, on which the other half of the die support frame will be mounted, is made in the same way as the movable plate, without machining the hole for holding the screw with trapezoidal thread, but having the same faces as the movable plate ground for a proper settlement.

The mold support frame will be made of aluminum, which serves two purposes: to maintain the outer shape of the mold made of plastic and to cool the mold after injecting the molten material.

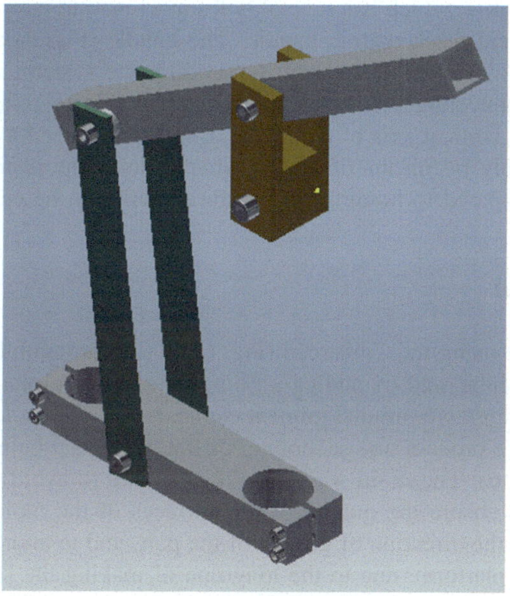

Fig. 5. Handle assembly

This part will also have a channel fitted at the top that fits on the profile of the injection nozzle, to ensure optimal injection, without leaks that could make it difficult to remove the injected part after it cools.

The injection assembly is based on a simple model, consisting of a piston body formed of a copper tube to have the best possible thermal conductivity, a nozzle for directing the material and a piston operated by a manual lever.

The heating cylinder was chosen to be made of a copper tube, for the good thermal qualities of the material. It has at one end a thread for mounting the nozzle adapter, necessary for attaching the injection nozzle. Its length of 200 mm together with the internal diameter of 40 mm allow a capacity of 313.75 g of molten material, with a volume of 251 cm^3, considering the density of PLA material (Polylactic Acid) of 1.25 g/cm^3.

The injection nozzle is made of brass, due to the good thermal conductivity of this material and due to the low coefficient of friction of this material. Being a low-cost material, the damage of this part can be solved by replacing the part with a new one, being easy to process. Its inner profile has the same idea of a conical profile to reduce the possibility of clogging due to obstructions such as foreign material such as dust that could accumulate in the sharp corners of this piece, which can lead to the nozzle blocking. The piston is made of easy-to-process parts, the piston arm being made of a square aluminum profile, which has several holes machined in it to allow the mounting of the piston head and the actuating handle. The handle (Fig. 5) also consists of the same aluminum profile as the piston arm to simplify the system, it is attached to the piston arm by a connecting piece that allows the handle to rotate around the axis perpendicular to the vertical axis of the piston arm. It is attached to the center body of the injection assembly by means of levers, which allows the piston to have a lower lowering stroke compared to holding the handle directly by the center body.

3 Experimental

The mold was made using the Tinkercad (Fig. 6) editing program by cutting the mold cavity from a solid cube, and by adding a filling gate, two air exhaust channels inside the mold and by adding two mold alignment elements [7]. The molds were made using the rapid prototyping process, the sectioning of the 3D model being done in the Cura 3.6.0 program (Fig. 6). The mold was made as a square to avoid the need to use the support material, to ensure the quality of the surfaces of the mold cavity, this being directed upwards in the direction of growth of the part, and to avoid detachment of the part from the work platform. due to the longtime of making the part and the temperature variation at the surface of the work platform. The molds were made using Rapid Prototyping SLS (Selective Laser Sintering) technology in plastic powder (Polyamide P2200), the equipment used FORMIGA P110 Velocis that uses a 30 W CO_2 laser, using a scanner system with a set of galvanometric mirrors, which gives it a printing speed of up to 5 m/s, and depending on the material used, the layer resolution of 0.10 mm [5].

Once the powder has been leveled, the laser driven by the galvanometric mirror system scans the surface of the powder by sintering the powder according to the layers of the workpiece (Fig. 7). They are fixed with a weak adhesive inside the model, on the whole device the mold being screwed inside the supports for a better fixation (Fig. 8). Generali view of the constructed device is presented in the Fig. 9.

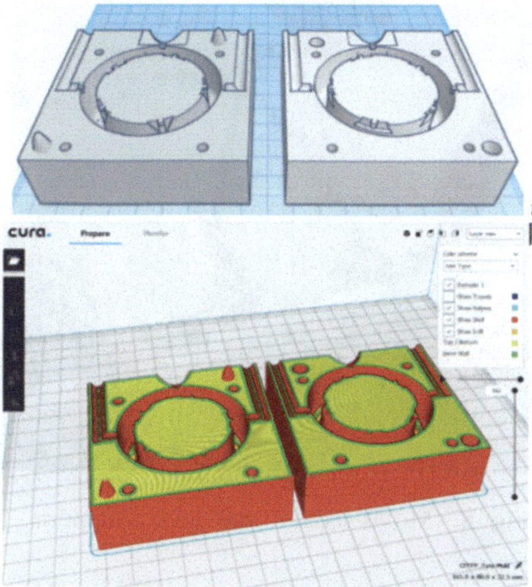

a

b

Fig. 6. Design and virtual processing of the molds between 3D printing: a - molds made using the Tinkercad editing program; b - preview of the molds in the CURA slicing program

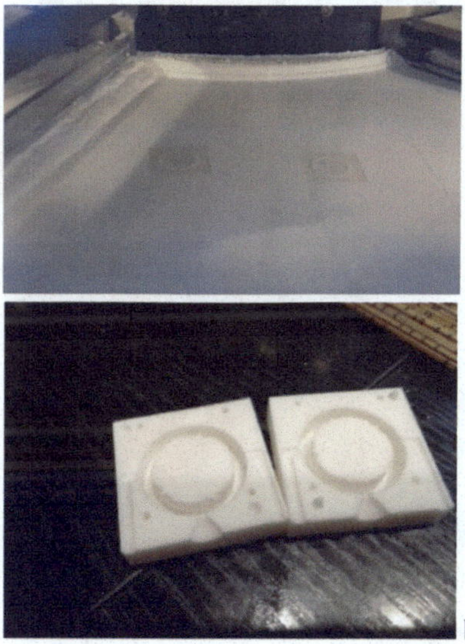

a

b

Fig. 7. Realizing of molds by selective laser sintering: a - Parts during manufacturing; b - mold made of powder raw material

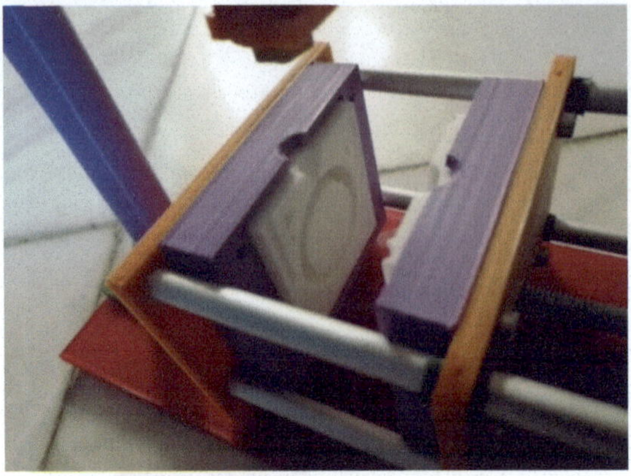

Fig. 8. Molds fixed in the device supports

Fig. 9. General view of the realized system

A Zhermack ZA 22 two-component rubber material with a tensile strength of 380% was used as the casting material [8]. The two components of the liquid rubber, the base and the catalyst, were mixed in equal proportions (a special scale was used). For optimal results, the mixing ratio must be strictly respected. After homogenizing the two components, the obtained material is injected to the model to be reproduced. The vulcanization of the flexible material was performed at room temperature (22 °C) in about 3 h, but a higher temperature reduces working time. Figure 10 presents the obtained results.

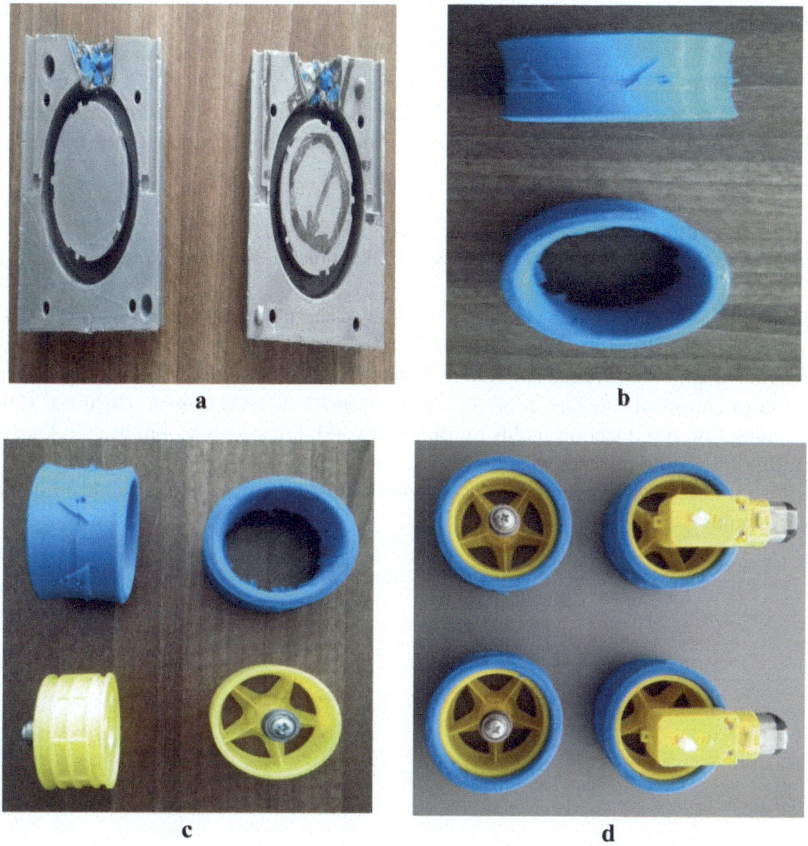

Fig. 10. Injection mold and product obtained after injection: a - molds; b - tire obtained by casting Zhermark material; c - tire and tire support; d - wheel with tire

4 Conclusions

The system designed and executed is very simple to make from a constructive point of view, and is also very easy to use, it is very easy to adjust and offers quick access to the molds made by rapid prototyping used for the injection process. This method can greatly reduce the costs of making small or medium series production parts, without sacrificing the quality of the parts produced. In addition to reducing the cost of manufacturing and the fabrication time is reduced several times in comparison with traditional technologies, and injection molds of any degree of complexity and any internal configuration (complex internal channels and cavities) can be created. These technologies can be easily integrated into the jewelry industry, in the production of dental and orthopedic products, in design bureaus, for research and development, in training and prototyping centers. Geometrically complex castings obtained as a result of the use of additive technologies can find application in cinema and on television, when it is required to quickly produce unusual props of complex shape.

References

1. Seres, I.: Matrite de injectat. Editura Imprimeriei de Vest, Oradea (1999)
2. Berce, P., Bâlc, N., Păcular, R., Brătean, S., Caizar, C., Radu, A.S., Fodorean, I.: Tehnologii de Fabricaţie prin Adăugare de Material şi Aplicaţiile Lor. Editura Academiei Române, Bucureşti (2013)
3. Berce P., Bâlc, N., Ancău, M., Comşa, S., Jidav, H., Caizar, C., Chezan, H.: Fabricarea Rapidă a Prototipurilor. Editura Tehnică, Bucureşti (2000)
4. Bâlc, N., Berce, P., Achimaş, G., Gyenge, G.: Tehnologii Neconvenţionale. Editura Dacia, Cluj-Napoca (2000)
5. Gheorghe, I., Moldovanu, A.: Microtehnologii avansate prin prototipare rapidă sinterizare cu laser. Editura CEFIN, Bucureşti (2010)
6. Năstase-Dan, C., Moldovanu, A.: Rapid Prototyping, Tehnologie avansată implementată în industria mecatronică. Lucrare conferinţă internaţională de mecatronică, Bucureşti (2005)
7. 3D Printing low-run injection molds by Ben Redwood. https://www.3dhubs.com/knowledge-base/3d-printing-low-run-injection-molds
8. https://www.zhermack.com/en/product_category/industrial/mould-making-industrial-en/master-mould/

Thermographic Analysis of 3D Printed Dental Models

Edgar Moraru(✉), Mariana-Florentina Stefanescu, Octavian Dontu, Ciprian Rizescu, and Carmen Draghici

University Politehnica of Bucharest, Splaiul Independentei no. 313, District 6, Bucharest, Romania
eddy_milan91@yahoo.com, octavdontu@yahoo.com, {mariana.stefanescu, ciprian.rizescu}@upb.ro, carmen.draghici3@gmail.com

Abstract. The paper aims to establish temperature ranges for additive technologies used to realize dental models. The problem of shrinkage and deformation at cooling of material for parts made by thermoplastic extrusion is well known, so it is necessary to research how the material cools after it has been manufactured using this technology. Polylactic acid was used as a raw material for the dental model and temperatures were measured in different areas of interest using a thermal imaging camera.

Keywords: Additive technologies · Dental models · Thermographic analysis

1 Introduction

Medicine, as a science, has always been closely linked to innovation and the latest advances in high technology. This is especially true for dentistry, an industry that is closely connected not only with high technology, but also very critical to the quality of the materials used. 3D printers occupy an increasingly important place in the work of any dental clinic, dental laboratory or research centers [1, 2]. With their help, dentists not only improve the quality of their products and services, but also save significant money. In addition, 3D printers in dentistry guarantee accelerated production volumes and incredible precision in finished products. 3D printers save dentists and dental technicians from a very complex and time-consuming process in work - manual modeling of prostheses, crowns and dental models. Customers no longer need to wait long and go through the entire complex process from the first visit to the installation of the final design, going through a series of fittings and improvements. Now they just need to do a scan of the oral cavity - and soon get an excellent result. The unique shape of each tooth of each pacient is incredibly difficult to obtain using manual manufacturing or a milling machine. But the main thing is that the dental models printed on a 3D printer exactly repeat all the nuances of the original sample [1, 2]. Fused deposition modeling technology [3] opens up great opportunities for dentists for realization of dental models: it can be stored all the anatomical data of patients in digital form, no more casts and samples - just print the desired model on a 3D printer equipment; significantly accelerate the production of the necessary models; fully automatic printing

G. I. Gheorghe (Ed.): ICOMECYME 2020, LNNS 143, pp. 149–155, 2020.
https://doi.org/10.1007/978-3-030-53973-3_16

process that eliminates the human factor. But still there is a known problem - the deformation and contraction of the printed model, which will result in errors in dimensional accuracy, and the study of cooling mode is very important. In this article it will be realised a dental model from a thermoplastic material – PLA (polylactic acid) and it will be studied on how it cools with the help of a camera with thermal imaging FLIR [4].

2 Materials and Methods

For the purposes of the article, a FLIR thermal imaging camera [4, 5] was used, which finds applications in detecting hidden problems due to imperceptible temperature rises. This thermal camera can also be useful for safety assessments, electrical fault avoidance, heating and cooling systems, automotive services and medicine [6]. Infrared heating information appears immediately on a color LCD screen and does not require prior experience with this equipment. The equipment used is shown in Fig. 1.

Fig. 1. Used infrared camera [4, 5]

Among the main technical characteristics of the device used are:

- Minimum focusing distance - 100 mm
- Imaging frequency - 9 Hz
- IR resolution – 140 × 140 pixels
- Measuring range - −20 °C–120 °C
- Margin of error - ±2 °C [4].

Images can also be downloaded to the computer and then uploaded to the ThermaCAM QuickReport software, which allows the analysis of images obtained on the infrared camera and their presentation in a report. The software allows the user to adjust the level, range, zoom and camera panel. The program also offers a wide range of different colors. Image information can be displayed to adjust the measuring instruments, object parameters, text comments, and the table of measurement results. To investigate how the material cools after the completion of the thermoplastic extrusion process and which often causes dimensional changes due to shrinkage, a PLA dental prosthetic model was made on the Wanhao Duplicator 4S printer (Fig. 2). PLA products hardens very quickly when using a fan for cooling. PLA minimally deforms when the temperature changes, including when it cools down after printing (other similar material in this domain ABS can become very deformed during uneven cooling). Also it is more environmentally friendly and safe than other materials, since annually renewable natural resources (for example, corn starch) are used for its synthesis. However in terms of mechanical properties polylactic acid is inferior to ABS (acrylonitrile butadiene styrene) [2, 7, 8].

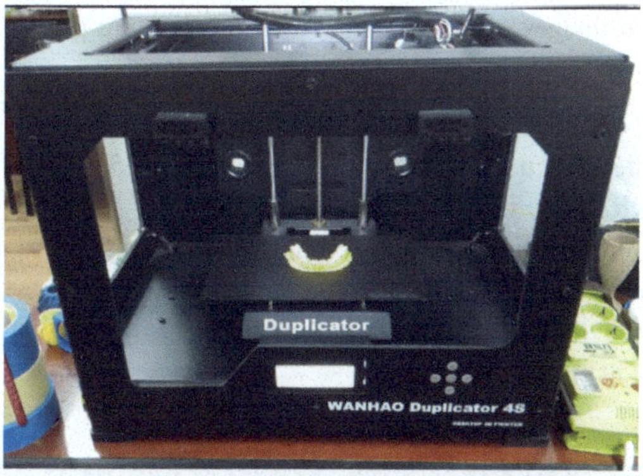

Fig. 2. Realised dental model on Wanhao Duplicator 4S 3D printer

3 Experimental

After the completion of the construction, it was waited until the entire thermal field will be in the area of the measuring range of the equipment used. Figure 3 shows thermal images of the polylactic acid prosthetic model after completion of construction, the difference between the images being about 30 s. It can be observed how the maximum temperature is found on the platform, which is usually heated up to 110 °C for optimal construction, and in terms of the dental model, higher temperatures are found in the molar area, this fact explicating by the larger volume than the incisor area and therefore a slower cooling. After printing, captures were taken for the dental model every 5 s for one minute to see how the material cools over time. From the "Analyze" menu of the program, three points of interest were set: the work platform, the central incisor area and the molar area (Fig. 4). The program allows the processing of captured data at any point of interest.

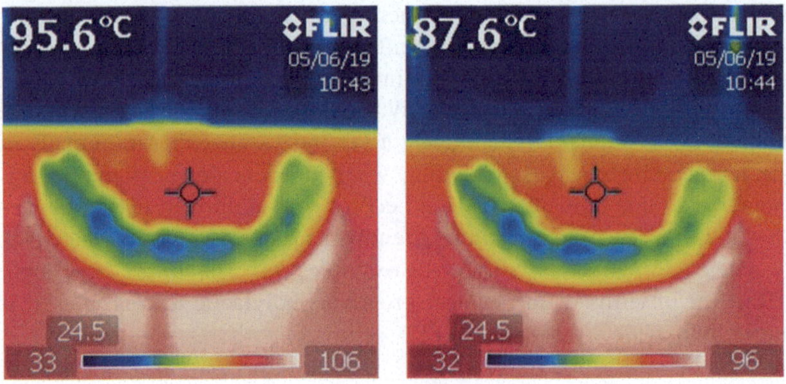

Fig. 3. Thermal images of dental model after the printing process is completed

Also, from the "Analyze" menu, a separate area can be defined, where the maximum and minimum temperature can be highlighted (Fig. 5). Other possibilities of the ThermaCAM QuickReport program include changing isotherms and temperature levels, making contours, areas or lines, creating a digital zoom up to 8 times, conducting an inspection report, highlighting the company, address and operator who performed the thermal analysis. In this report the parameters of the image and the investigated model can also be passed.

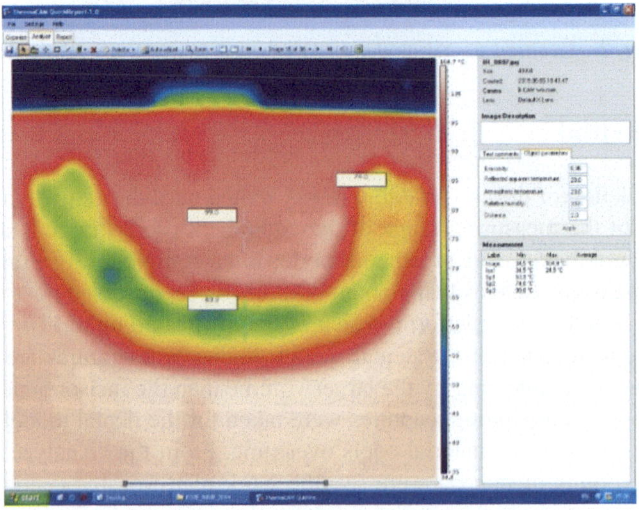

Fig. 4. Defining the points of interest in the program

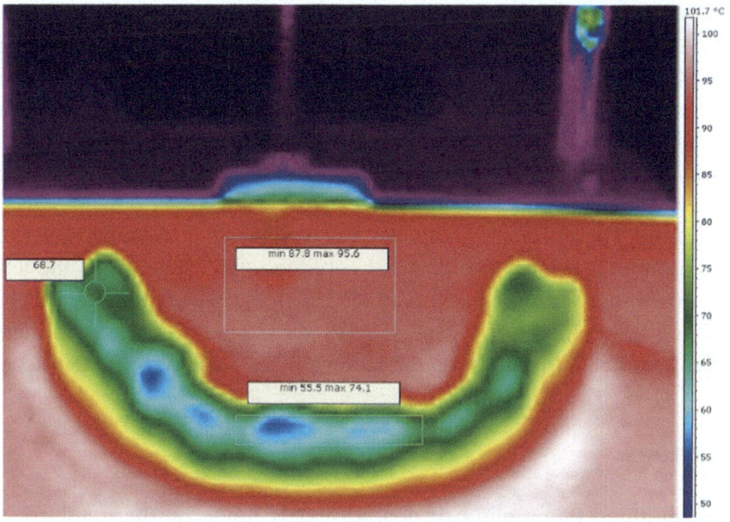

Fig. 5. Defining areas of interest in the program

4 Results and Discussions

Based on the thermal images obtained and processed in the ThermaCAM QuickReport software, it obtained the following cooling curves over time of the three areas of interest investigated, which are presented in Fig. 6.

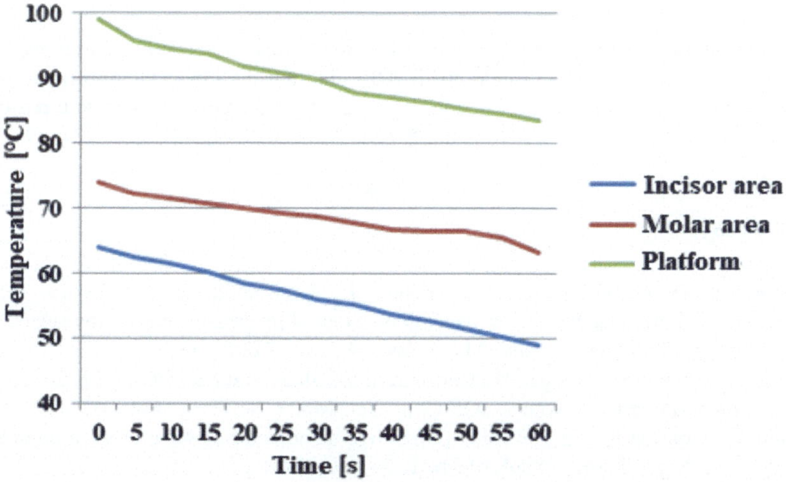

Fig. 6. Temperature as a function of time in different areas of the model

It can be seen from the graph how in one minute the temperature of the platform decreases by 15.5 °C, the temperature in the molar area by 11.1 °C, and in the incisor area by 14.8 °C. There is a sudden and uneven cooling that causes contraction, or even deformation of the printed part. Some thermoplastic materials are amorphous in nature. When the material heats up, the polymer chains move away from each other and the forces that bind them become weaker. In this case the material increases its volume and reduces its density. But when it cools after extrusion, the process begins to run in the opposite direction. The polymer chains approach each other, and the material shrinks. This is more pronounced when the material cools too quickly or when the temperature is unevenly distributed around the model. Even if the shrinkage rate of the studied material is much lower than in the case of ABS shrinkage, it is very important that the cooling of the material is optimal in order to obtain the designed characteristics [9–11].

5 Conclusions

The paper highlighted the cooling of dental models made by thermoplastic extrusion that can have a lot of beneficial applications in the field of dental restorative prosthetics. A solution to avoid shrinkage and deformation of the model after cooling is to maintain the cooling rate in the optimal range of material used, but this is difficult to achieve in an open 3D printer where many disturbing factors can influence the cooling rate. A closed-chamber printer and a heated platform can be used to solve this problem and effectively reduce material shrinkage. The model will be gradually cooled after printing is completed in closed and controlled conditions. In this way, all layers are cooled simultaneously at the same rate, and the temperature is uniform over the entire construction surface to prevent possible unwanted deformations. Future studies aim at a more detailed study of several materials and technologies used.

Acknowledgments. This work has been funded by the European Social Fund from the Sectoral Operational Programme Human Capital 2014–2020, through the Financial Agreement with the title "Scholarships for entrepreneurial education among doctoral students and postdoctoral researchers (Be Antreprenor!)", Contract no. 51680/09.07.2019 - SMIS code: 124539.

References

1. Moraru, E., Dontu, O., Petre, A., Vaireanu, D., Constantinescu, F., Besnea, D.: Some technological particularities on the execution of dental prostheses realized by selective laser deposition. J. Optoelectron. Adv. Mater. **20**(3–4), 208–213 (2018)
2. Moraru, E., Dontu, O., Besnea, D., Constantin, V.: Study and realization of prosthetic dental models by additive technologies. IOP Conf. Ser. Mater. Sci. Eng. **444**, 042017 (2018)
3. Spanu, A., Constantin, V.: The using of additive manufacturing for prototype production of moulds. Int. J. Mechatron. Appl. Mech. **1**, 7–11 (2017)
4. http://www.farnell.com/datasheets/1718881.pdf
5. https://www.flir.com/
6. Rizescu, D., Rizescu, C.: Temperature gradient analysis by thermography used in optometry. Int. J. Mechatron. Appl. Mech. **6**(II), 7–14 (2019)

7. Ramesh, M., Panneerselvam, K.: PLA-based material design and investigation of its properties by FDM. In: Shunmugam, M.S., Kanthababu, M. (eds.) Advances in Additive Manufacturing and Joining. LNMIE, pp. 229–241. Springer, Singapore (2020). https://doi.org/10.1007/978-981-32-9433-2_20
8. Chacon, J.M., Caminero, M.A., Garcia-Plaza, E., Nunez, P.J.: Additive manufacturing of PLA structures using fused deposition modelling: effect of process parameters on mechanical properties and their optimal selection. Mater. Des. **124**, 143–157 (2017)
9. Choi, Y., Kim, C., Jeong, H., Youn, J.: Influence of bed temperature on heat shrinkage shape error in FDM additive manufacturing of the ABS-engineering plastic. World J. Eng. Technol. **4**, 186–192 (2016)
10. Hrituc, A., Slatineanu, L., Mihalache, A., Dodun, O., Coteata, M., Nagit, G.: Accuracy of polylactide parts made by 3D printing. Macromol. Symp. **389**, 1900064 (2020)
11. Besnea, D., Spanu, A., Contantin, V.: The main factors affecting the accuracy during the manufacturing process based on material extrusion. Int. J. Mechatron. Appl. Mech. **2**, 59–65 (2017)

A Review in Biomechanics Modeling

Andreea-Mihaela Let[1], Viviana Filip[1,2(✉)], Dorin Let[2], and Simona Mihai[2]

[1] Doctoral School, Valahia University of Targoviste, 130004 Targoviste, DB, Romania
v_filip@yahoo.com
[2] Institute of Multidisciplinary Research for Science and Technology, Valahia University of Targoviste, 130004 Targoviste, DB, Romania

Abstract. Human kinetics are of major medical interest especially in case of disabled persons. Rehabilitation physicians and patients can be aided by bioinformatics and modeling techniques into tailoring their efforts. In this paper we are going to review computer simulation environments and sensor measurements in motion dynamics; them being used in rehabilitation procedures.

Keywords: Biomechanics · Human kinetics · Physical medicine and rehabilitation

1 Introduction

[A Deep Learning Framework for Assessing Physical Rehabilitation Exercises]

Taking part in a physiotherapy and rehabilitation program is regularly mandatory in postoperative recovery. Although it is usually not reasonably economic offering patient access to a clinician for every single rehabilitation session [1]. Therefore, common practice in healthcare is for the initial part of rehabilitation programs to be carried out in a stationary facility under direct supervision by the clinician and then the second part to be performed in outpatient conditions where patients perform a set of prescribed exercises in their own home. Literature study show that more than 90% of all rehabilitation sessions are conducted at home [2]. Considering this, patients are usually asked to record their daily progress and visit the clinic afterwards for functional assessment. Nevertheless, many medical resources report low levels of patient dependence on the proposed exercise regimes in home-based rehabilitation, leading to long-term treatment times and increased health costs [3, 4]. While various factors have been identified that contribute to a reduction in compliance, the main impact factor is lack of ongoing feedback and monitoring of patient activity by healthcare providers [5]. Although a variety of tools and devices have been developed to support physical rehabilitation, for example: KINECT-based assistance, robotic assistive systems [6], VR and game interfaces [7] there is still a shortage of useful, powerful systems for automatic monitoring and evaluation of patient performance.

A computer-based assessment of physical rehabilitation requires a patient's performance assessment to complete the prescribed rehabilitation exercises, based on the sensory motion data processing. Although the primary role of rehabilitation assessment

G. I. Gheorghe (Ed.): ICOMECYME 2020, LNNS 143, pp. 156–164, 2020.
https://doi.org/10.1007/978-3-030-53973-3_17

is to improve patient results and lower health-care costs, current methods lack diversity, strength, and practical importance [8].

2 Physical Medicine and Rehabilitation

Natural Medicine and rehabilitation, also known as physiatry medicine or rehabilitation medicine, intends to augment and restore functional capacity and quality of life for those with physical disabilities or disabilities affecting the spinal cord, nerves, brain, bones, joints, ligaments, and tendons [9]. A doctor who has completed the training in this area is referred to as a physiatrist. Distinct from other medical disciplines that focus on medical "therapy", the objectives of natural therapy are to maximize the independence of patients in daily life activities and improve the quality of life.

This branch of medicine deals with the prevention, diagnosis, treatment, and rehabilitation management in case of patients with impaired medical conditions and is divided into subspecialties for neurorehabilitation, amputation of limbs, heart injury, spinal column, trauma, and general rehabilitation [10]. Main target groups are young people and those of working age, but aspects of specialization, particularly in relation to technical assistance, orthopedic or prosthetic orthopedics, wheelchair provision, are relevant to persons of all ages. The main objectives are to identify deficiencies that restrict daily activity and tasks; improve physical and cognitive performance; and modify personal and environmental factors to facilitate partaking and quality of life.

2.1 Deficiency Rehabilitation and Prosthetics

In medicine, a deficiency is a lack or shortage of a functional entity, by less than normal or necessary supply or function. In this paper the focus will be over motor functions.

The efficiency of gait/walking is equal to the cost of metabolism (VO, in R &/kg per m) split by the mechanical work required to produce movement [11]. In human gait, the main difficulty in calculating efficiency was to calculate the mechanical work performed on body parts [12]. The mechanical operation of human gait mainly concerns three functions: Support of the body's mass during the standstill phase, acceleration, and deceleration, raising and lowering the body's center of mass and slowing of the lower extremities during acceleration and oscillation phase. The main determinant of the metabolic cost of gait is usually devoted to the study of moving the body's center of mass [13]. Biomechanical factors that contribute to the increased metabolic cost of gait in pathological walking due to neurological disorders, musculoskeletal disorders or amputations have not been well illuminated. Excessive co-contraction, abnormal acceleration and deceleration of body and extremities, and abnormal transfer of energy are usually studied in these matters [14].

2.2 Musculoskeletal Rehabilitation

Muscle disorders associated with knee osteoarthritis are the root cause behind functional limitations. It is very important to demonstrate the scope of muscle disorders, the relationship with physical function and disease progression, and the back-fitness therapy that targets muscle disorders. Patients with knee osteoarthritis have significant muscle disorders [15]. These muscle disorders affect the physical function and therapy should be targeted. More study is needed to investigate the relationship between quadriceps force and knee osteoarthritis start and progression, and to determine optimal pre-exercise coding that improves results in this patient population [16].

The purpose of knee rehabilitation is bi-directional. The first is to prevent the muscles surrounding the knee from weakening. The second is to reduce the load on the knee joint. People with stronger muscles around the knee usually have fewer problems with the joint. Weaker muscles provide less support, creating more work for the knee joint [17]. Contrarywise, the strong muscles of the leg enhance support and control of the knee joint.

3 Biomechanics and Bioinformatics

Biomechanics is a fundamental science used by orthopedic surgeons, neurosurgeons, osteopaths, physiatrists, rheumatologists, physical and occupational therapists, chiropractors, athletic trainers and beyond. Biology is the study of living things; mechanics is the study of motions and the applied loads that cause them. Biomechanics can be defined, therefore, as the study of the motions experienced by living things in response to applied loads [18]. A secondary definition could be implementation of the mechanics to better understand the impact of the applied loads on the structure and the features and functions of the structures they interact with. Therefore, the field of biomechanics is extensive. It can also include studying the effects of the mechanical properties of food, wind load or gravity on plant growth, the flight of birds, the drag reduction properties of dolphin's skin [19]. In addition, biomechanics solves many health problems, including human, animal diseases, injuries, and their treatment. Biomechanics is the foundation of the rapidly developing field of biomedical engineering.

3.1 Bioinformatics

Bioinformatics is an interdisciplinary field that has developed methods and software tools for understanding biological data, especially in the case of large and complex data sets. Bioinformatics combines information engineering, computer science, mathematics, biology, and statistics to analyze and interpret biological data. It is usually being used for computer analysis of biological queries using mathematical and statistical techniques. It includes biological studies that use computer programming in addition to the new methodology, as well as a specific "channelization" that is used repeatedly, particularly in the field of genomics. In a more formal way, bioinformatics also tries to understand the principles of the organization of the nucleus and protections, anonymous proteomics [20]. Initially, many bioinformatics researches targeted a relatively narrow focus, focusing on creating algorithms to analyze specific types of data, such as gene sequences or protein structures. Nowadays the goals of bioinformatics are integrated and aim to find out how combinations of different types of data can be used to understand natural phenomena.

3.2 Human Movement Modeling

Traditional approaches to mathematical modeling of human movements are generally divided into two categories: top-down approaches that promote hidden states to define the temporal dynamics of movements and bottom-up approaches that use local resources to represent movements. Methods commonly used in the first category include Kalman filters [21], hidden Markov models [22] and Gaussian mixing models [23]. The main deficiencies of these methods originate from the use of linear models for the transitions between latent states (as in Kalman filters) or from the adoption of simple internal structures of latent states (typical for hidden Markov models). Approaches based on the extraction of local characteristics employ predefined criteria to identify key points [24] and/or necessary body postures [25] or a collection of movement statistics (e.g., mean, standard deviation, mode, median) [26]. Local features are typically motion-specific, this limits the possibility of efficiently implementing spatial-temporal variations in a movement model.

3.3 Movement Assessment

The quantification of the level of correction at the conclusion of prescribed exercises is important for the development of tools and devices to support home rehabilitation. The evaluation of movement in existing studies is usually performed by comparing a patient's performance of an exercise with the performance desired by healthy participants.

Several studies in the literature on exercise evaluation have used machine learning methods to classify individual repetitions into correct or incorrect motion classes. The methods used for this purpose include the Adaboost classifier, the closest neighbors k, the Bayesian classifier, and a set of multi-layered perceptron NNs [27]. The outputs in these approaches are discrete class values of 0 or 1 (that is, incorrect or correct repetition). However, these methods do not provide the ability to detect variable levels of motion quality or identify incremental changes in patient performance during the rehabilitation program.

Most related studies employ distance functions to obtain motion quality indices. Specifically, used a variant of mahalanobis distance to quantify the level of correction of rehabilitation movements, based on a calculated distance between the repetitions performed by the patient and a set of repetitions performed by a group of healthy individuals [28]. Similarly, a body of work used the dynamic time distortion algorithm (DTW) [29] to calculate the distance between a patient's performance and the performance of healthy individuals [30]. The advantage of distance functions is that they are not specific to the exercise and therefore can be applied in the evaluation of new types of exercises. Distance functions also have deficiencies because they do not attempt to derive a model from rehabilitation data, and distances are calculated at the level of individual time intervals in gross sensory measurements.

Several researches used probabilistic approaches for modelling and evaluating rehabilitation movements. Studies based on hidden Markov models [31], and mixtures of Gaussian distributions [32] usually carry out quality assessments based on the likelihood that individual sequences are drawn from the trained model. Although the use of probabilistic models is advantageous in addressing the variability caused by the stochastic nature of human movements, models with capabilities for hierarchical data representation can produce more reliable results to evaluate the quality of movement and better generalize new exercises.

4 Biomechanics Modeling

Highly coordinated actions from our musculoskeletal system are required even "simple" tasks like catching an object. Such coordinated movements are mainly operated by voluntary contractions of skeletal muscles. The underlying mechanism of employment and muscle power production is a challenge and subject to a lot of research [33].

A non-invasive and clinically available diagnostic tool used to obtain neuromuscular system operation (or dysfunction) is electromyographic (EMG); measurement of the potential induced over skin surface, caused by muscle activation [34]. Results on neuromuscular system are often subtracted from those obtained through signal processing, but these signal processing techniques typically ignore underlying muscle structure. Other limitations of EMG measurements are, for example, capturing activity from muscle parts close to the surface only. This leads to difficulties in identifying, for example, crosstalk [35]. Furthermore, an EMG often records weak signals due to layers of fat tissue and is sometimes limited to isometric contractions.

Therefore, calculation models can be used to obtain a more complete insight into the neuromuscular system [36]. These models need to capture most of the electromechanical properties of skeletal muscle tissue and the interaction between nerve intake and muscle contraction. The contraction behavior of the skeleton muscle tissue is heavily modeled using lumped parameter models, such as hill-type skeletal muscle models [37], continuum mechanical skeletal muscle models [38], or multi-scale, chemo-electromechanical skeletal muscle models [39]. Analytical models are used to predict the resulting EMG of a particular stimulation [40], or numerical approaches [41]. However, for realistic muscle geometries, numerical methods are almost inevitable. Retouching and rear. The chemo-electromechanical models [42], are especially suitable for including many structural and functional characteristics of skeletal muscles. The one-dimensional calculation embeds muscle fibers into a three-dimensional frame muscle model and associates them with a specific motor unit. Furthermore, these models can be directly linked to motor neuron models either phenomenologically [43] or biophysically [44] to further investigate the relationship between neural and mechanical behavior. The degree of detail and complexity obtained within these models requires that different physical phenomena to be combined on different temporal and spatial scales, for example, mechanical or electrical condition of muscle tissue on organ scale and the bio-chemical processes on the cellular scale.

Flexible multi-scale, multi-physics calculation framework and significant computational power are required to account for all these different processes at different scales. Computational resources limit the number of individual muscle fibers that can be thought of within the skeletal muscle model. International open source libraries containing chemo-electromechanical models like OpenCMISS [45] allow general muscle geometries with thousands embedded computational muscle fibers. However, because most skeletal muscles have significantly more fibers [46], the embedded muscle fibers represent a choice of geometrically only real muscle fibers in their geometric surroundings. Simulations containing less than a thousand fibers can potentially provide some information about the neuromuscular system, while some effects on all motor units and muscle fibers, such as the recruitment of the motor unit and the effect on the resulting EMG, cannot be estimated. Simulating a detailed and realistic model with a realistic number of muscle fibers allows us to estimate the accuracy of the "reduced" models by comparing them to the output of the detailed full "benchmark" model.

The skeleton muscle tissue and heart muscle tissue reshare many similarities in the spectrum with its underlying microstructure. Therefore, similar simulation techniques can be used for both heart models and skeletal muscle models. However, there are significant differences in the spread of recruitment and action potential between the heart and skeletal muscle tissue [47]. Although the action potential is homogeneous and continuously spread across a three-dimensional myocardium, the behavior of the skeletal muscle is a radiant spread of the potential for heterogeneous recruitment and action - in fact, each muscle fiber can lead to intricate potential areas independently [48].

The majority of multi-functional computational products for biomedical applications like OpenCMISS, are developed to provide flexibility in the use of simulation tools for them, but they are not normally designed for highly parallel simulations on supercomputers. This can be accomplished using standards: CellML [49] and FieldML [41]. The standards are used to improve existing models of physical physics for a wide range of applications.

The majority of calculations are designed to be run on small computational clusters. Even if typically, they can be compiled on large scale HPC computing clusters [50], they often cannot take full advantage of the hardware's potential for several reasons. In addition, simulation run time is typically considered to be less important than model complexity and output. Therefore, typical simulations of biomedical applications are not usually optimized for numerical efficiency, parallel scalability, the use of new algorithms, or file I/O [51]. After large-scale simulations of biomedical applications are resolved in high detail, the most specialized visualization tools, such as OpenCMISS-Zinc [52], can no longer process large amounts of simulation data.

5 Conclusions

Numerous researchers have been concerned with the development of biomechanics modeling programs (e.g. Anybody, Cosmos Motion, 3D Motion Analysis, Motion Analysis Software) in order to study the mechanics of the human body in virtual space. Numerical simulations performed with the help of these programs help us to obtain important data on the kinematics (speeds, accelerations) and dynamics (forces and moments of reaction) to which the joints of the human body are subjected. When these demands are high, they can lead to joint disease and later to the need to replace them. Therefore, knowing the behavior over time of the joints subjected to natural mechanical stress is particularly useful, in order to avoid or postpone surgical operations to prosthetic joints. Full model computational power needed is increasing alongside our knowledge and need for better visualization. New technologies like artificial intelligence Ai and machine learning ML can further enhance bioengineering computation, but this involves a totally novel approach.

References

1. Plant, S., Tyson, S.F.: A multicentre study of how goal-setting is practised during inpatient stroke rehabilitation. Clin. Rehabil. **32**(2), 263–272 (2018)
2. Swartz, M.C., et al.: Effect of home-and community-based physical activity interventions on physical function among cancer survivors: a systematic review and meta-analysis. Arch. Phys. Med. Rehabil. **98**(8), 1652–1665 (2017)
3. Hoaas, H., et al.: Adherence and factors affecting satisfaction in long-term telerehabilitation for patients with chronic obstructive pulmonary disease: a mixed methods study. BMC Med. Inform. Decis. making **16**(1), 26 (2016)
4. Fortis, P., et al.: A home-based prism adaptation training for neglect patients. Cortex **122**, 61–80 (2018)
5. Del Boca, F.K., et al.: The SBIRT program matrix: a conceptual framework for program implementation and evaluation. Addiction **112**, 12–22 (2017)
6. Liao, Y.-T., et al.: Development of kinect-based upper-limb assistance device for the motions of activities of daily living. In: 2018 IEEE International Conference on Cyborg and Bionic Systems (CBS). IEEE (2018)
7. Trombetta, M., et al.: Motion Rehab AVE 3D: a VR-based exergame for post-stroke rehabilitation. Comput. Methods Programs Biomed. **151**, 15–20 (2017)
8. Gustavsson, C., Jessica, N., Anne, S.: Activity and life-role targeting rehabilitation for persistent pain: feasibility of an intervention in primary healthcare. Eur. J. Physiotherapy **20** (3), 141–151 (2018)
9. Balaşa, M.-C., Filip, V.: Analysis of the static and dynamic mechanical behavior of a tibial bone-knee implant assembly without a tibial extension. In: Gheorghe, G.I. (ed.) ICOMECYME 2019. LNNS, vol. 85, pp. 194–205. Springer, Cham (2020). https://doi.org/10.1007/978-3-030-26991-3_18
10. Balasa, M.C., Cuculici, S., Filip V.: Modelling of a tibial bone-knee implant assembly in order to analyse its mechanical behaviour. In: Proceeding of International Joint Conference on Materials Science and Mechanical Engineering, Materials Science and Engineering (2020). https://doi.org/10.1088/1757-899x/715/1/012019
11. Koelewijn, A.D., Heinrich, D., Van Den Bogert, A.J.: Metabolic cost calculations of gait using musculoskeletal energy models, a comparison study. PLoS ONE **14**(9), e0222037 (2019)
12. Mihai, S., Filip, S., Vladescu, M.: Contributions to the improvement of the tribological behaviour of hip implant joints. J. Sci. Arts **16**(35), 177–184 (2016)
13. Balasa, M.C., et al.: Using 3D scanning techniques in orthopedic systems modeling. Sci. Bull. Valahia Univ. Mater. Mech. **15**(13), 41–47 (2017). https://doi.org/10.1515/bsmm-2017-0016
14. Florescu, V., Capitanu, L., Badita, L., Filip, V.: A novel engineering spherical bearing, with potential application for a hip implant. J. Mech. Eng. Autom. (JMEA) **6**(5), 217–226 (2016). https://doi.org/10.17265/2159-5275/2016.05.001
15. Balaşa, M.-C., Mihai, S., Filip, V., Negrea, A.-D., Tomescu, G.: Modelling the tibial bone using cad techniques, starting from the 3D scan model. In: Gheorghe, G.I. (ed.) ICOMECYME 2018. LNNS, vol. 48, pp. 156–165. Springer, Cham (2019). https://doi.org/10.1007/978-3-319-96358-7_16

16. Florescu, V., Capitanu, L., Badita, L., Filip, V.: A surprising tribological validation: metal on metal total hip prosthesis with rolling friction. Int. J. Appl. Innov. Eng. Manage. **5–7**, 025–038 (2016)
17. Nur, H., Sertkaya, B.S., Tuncer, T.: Determinants of physical functioning in women with knee osteoarthritis. Aging Clin. Exp. Res. **30**(4), 299–306 (2018)
18. Lim, Y.P., Lin, Y.-C., Pandy, M.G.: Effects of step length and step frequency on lower-limb muscle function in human gait. J. Biomech. **57**, 1–7 (2017)
19. Cheng, R., et al.: Biomechanics of human motion. In: Frontiers in Orthopaedic Biomechanics, pp. 265–300. Springer, Singapore (2020)
20. Bludau, I., Aebersold, R.: Proteomic and interactomic insights into the molecular basis of cell functional diversity. Nat. Rev. Mol. Cell Biol. **21**, 1–14 (2020)
21. Joukov, V., et al.: Rhythmic extended Kalman filter for gait rehabilitation motion estimation and segmentation. IEEE Trans. Neural Syst. Rehabil. Eng. **26**(2), 407–418 (2017)
22. Ma, H., Liao, W.-H.: Human gait modeling and analysis using a semi-Markov process with ground reaction forces. IEEE Trans. Neural Syst. Rehabil. Eng. **25**(6), 597–607 (2016)
23. Dolatabadi, E., et al.: Mixture-model clustering of pathological gait patterns. IEEE J. Biomed. Health Inform. **21**(5), 1297–1305 (2016)
24. Papavasileiou, I., Zhang, W., Han, S.: Real-time data-driven gait phase detection using infinite gaussian mixture model and parallel particle filter. In: 2016 IEEE First International Conference on Connected Health: Applications, Systems and Engineering Technologies (CHASE). IEEE (2016)
25. Prakash, C., Sujil, A., Kumar, R., Mittal, N.: Linear prediction model for joint movement of lower extremity. In: Sa, P.K., Bakshi, S., Hatzilygeroudis, Ioannis K., Sahoo, M.N. (eds.) Recent Findings in Intelligent Computing Techniques. AISC, vol. 707, pp. 235–243. Springer, Singapore (2019). https://doi.org/10.1007/978-981-10-8639-7_24
26. Cheng, W., et al.: Inertial sensors to detect multiple gait disorders. In: 2019 IEEE International Conference on Dependable, Autonomic and Secure Computing, International Conference on Pervasive Intelligence and Computing, International Conference on Cloud and Big Data Computing, International Conference on Cyber Science and Technology Congress (DASC/PiCom/CBDCom/CyberSciTech). IEEE (2019)
27. Barra, P., Bisogni, C., Nappi, M., Freire-Obregón, D., Castrillón-Santana, M.: Gait analysis for gender classification in forensics. In: Wang, G., Bhuiyan, M.Z.A., De Capitani di Vimercati, S., Ren, Y. (eds.) DependSys 2019. CCIS, vol. 1123, pp. 180–190. Springer, Singapore (2019). https://doi.org/10.1007/978-981-15-1304-6_15
28. Leys, C., et al.: Detecting multivariate outliers: use a robust variant of the Mahalanobis distance. J. Exp. Soc. Psychol. **74**, 150–156 (2018)
29. Li, M., et al.: Research of gait recognition based on human electrostatic signal. In: 2018 2nd IEEE Advanced Information Management, Communicates, Electronic and Automation Control Conference (IMCEC). IEEE (2018)
30. Das, S., Meher, S.: Analysis of spatio-temporal dynamic patterns of gait for recognition (2017)
31. Rybarczyk, Y., et al.: Implementation and assessment of an intelligent motor tele-rehabilitation platform. Electronics **8**(1), 58 (2019)
32. Kang, K.-T., et al.: Probabilistic evaluation of the material properties of the in vivo subject-specific articular surface using a computational model. J. Biomed. Mater. Res. Part B Appl. Biomater. **105**(6), 1390–1400 (2017)
33. Teklemariam, A., Hodson-Tole, E., Reeves, N.D., Cooper, G.: A micromechanical muscle model for determining the impact of motor unit fiber clustering on force transmission in aging skeletal muscle. Biomech. Model. Mechanobiol. **18**(5), 1401–1413 (2019). https://doi.org/10.1007/s10237-019-01152-2

34. Wang, J., et al.: A Static Distributed-parameter Circuit Model Explains Electrical Stimulation on the Neuromuscular System. arXiv preprint arXiv:1812.00598 (2018)
35. Mesin, L.: Crosstalk in surface electromyogram: literature review and some insights. Phys. Eng. Sci. Med. (2020)
36. Eskov, V.V., et al.: Chaotic dynamics of neuromuscular system parameters and the problems of the evolution of complexity. Biophysics **62**(6), 961–966 (2017)
37. Schmitt, S., Günther, M., Häufle, D.F.B.: The dynamics of the skeletal muscle: a systems biophysics perspective on muscle modeling with the focus on Hill-type muscle models. GAMM-Mitteilungen **42**(3), e201900013 (2019)
38. Bleiler, C., Castañeda, P.P., Roehrle, O.: A microstructurally-based, multi-scale, continuum-mechanical model for the passive behaviour of skeletal muscle tissue. J. Mech. Behav. Biomed. Mater. **97**, 171–186 (2019)
39. Bi, L., Guan, C.: A review on EMG-based motor intention prediction of continuous human upper limb motion for human-robot collaboration. Biomed. Sig. Process. Control **51**, 113–127 (2019)
40. Heine, C.B., Menegaldo, L.L.: Numerical validation of a subject-specific parameter identification approach of a quadriceps femoris EMG-driven model. Med. Eng. Phys. **53**, 66–74 (2018)
41. Fernandez, J., et al.: Musculoskeletal modelling and the physiome project. In: Pivonka, P. (ed.) Multiscale Mechanobiology of Bone Remodeling and Adaptation. CICMS, vol. 578, pp. 123–174. Springer, Cham (2018). https://doi.org/10.1007/978-3-319-58845-2_3
42. Röhrle, O., et al.: Multiscale modeling of the neuromuscular system: coupling neurophysiology and skeletal muscle mechanics. Wiley Interdisc. Rev. Syst. Biol. Med. **11**(6), e1457 (2019)
43. Ting, L,H., Chiel, H.J.: Muscle, biomechanics, and implications for neural control. Neurobiol. Mot. Control Fundam. Concepts New Dir., 365–416 (2017)
44. Schmid, L., et al.: Characterization of electromechanical delay based on a biophysical multi-scale skeletal muscle model. Front. Physiol. **10**, 1270 (2019)
45. Wittek, A., et al.: From finite element meshes to clouds of points: a review of methods for generation of computational biomechanics models for patient-specific applications. Ann. Biomed. Eng. **44**(1), 3–15 (2016)
46. Dao, T.T., Tho, M.-C.H.B.: A systematic review of continuum modeling of skeletal muscles: current trends, limitations, and recommendations. Appl. Bionics Biomech. **2018**, 17 (2018)
47. Di Cicco, E., et al.: Heart and skeletal muscle inflammation (HSMI) disease diagnosed on a British Columbia salmon farm through a longitudinal farm study. PLoS ONE **12**(2), e0171471 (2017)
48. Sandberg, C., et al.: Slower skeletal muscle oxygenation kinetics in adults with complex congenital heart disease. Can. J. Cardiol. **35**(12), 1815–1823 (2019)
49. Daoud, M.S., Ghadi, Y., Almimi, H.: Optimization of the application software in biomechanics and their contribution to the biological field. J. Eng. Sci. Technol. Rev. **12** (1), 178–184 (2019)
50. Tavares, João Manuel R.S., Dey, N., Joshi, A. (eds.): BIOCOM 2018. LNCVB, vol. 32. Springer, Cham (2020). https://doi.org/10.1007/978-3-030-21726-6
51. Kok, M., et al.: A scalable and distributed solution to the inertial motion capture problem. In: 2016 19th International Conference on Information Fusion (FUSION). IEEE (2016)
52. Bradley, C., et al.: Towards realistic HPC models of the neuromuscular system. arXiv preprint arXiv:1802.03211 (2018)

Intelligent Devices for Transporting Parts Between Processing Systems, Ultra-precise Cyber-Mechatronic Systems for Industrial and Laboratory Control (For Molded Parts in the Automotive Industry) and the Assembly Line

Badea Sorin-Ionut[1,2(✉)]

[1] Doctoral School, Valahia University of Targoviste, Târgovişte, Romania
sorin_ib@yahoo.com
[2] INCDMTM, 6-8 Pantelimon Road, 2nd District, 021631 Bucharest, Romania

Abstract. This article presents technical solutions for transporting parts, inside an automotive production plant, from processing equipment to mechatronic measurement and control systems and further to the assembly line or storage areas. These solutions help to implement Industry 4.0 in production plants by relieving factory personnel of certain routine tasks.

Keywords: Autonomous · Cobots · Industry 4.0

1 Introduction

Mobile robots accelerate the flow of materials to workstations and between manufacturing processes, helping to consolidate storage space and future operations. Robotic mobility leads to traceability and predictability on the road to Industry 4.0 and is crucial for faster development of new products.

Today, robots cost less and are easier to use. While the current adoption rate is about 40%, the adoption rate of robotics and automation is expected to increase to 80% in the next 10 years. For many manufacturers, finding enough skilled workers is a major challenge and one of the main factors in the development of robots.

Automated and autonomous words are easily confused and often changed, but their meaning is different.

Automated: Computer-controlled machines that can perform a set of defined tasks following specific instructions with a minimum or even no human intervention.

Autonomous: Machines that make decisions when faced with new unexpected situations. These machines have the ability to learn as they encounter new situations.

AGV-Automated Guided Vehicles and SDV-Self-Driving Vehicles' (or AMR-Autonomous Mobile Robots) are also often confused.

G. I. Gheorghe (Ed.): ICOMECYME 2020, LNNS 143, pp. 165–172, 2020.
https://doi.org/10.1007/978-3-030-53973-3_18

Each system works with a fundamentally different technology, from perception and navigation software to on-board sensors. Therefore, they have different capabilities and potential applications.

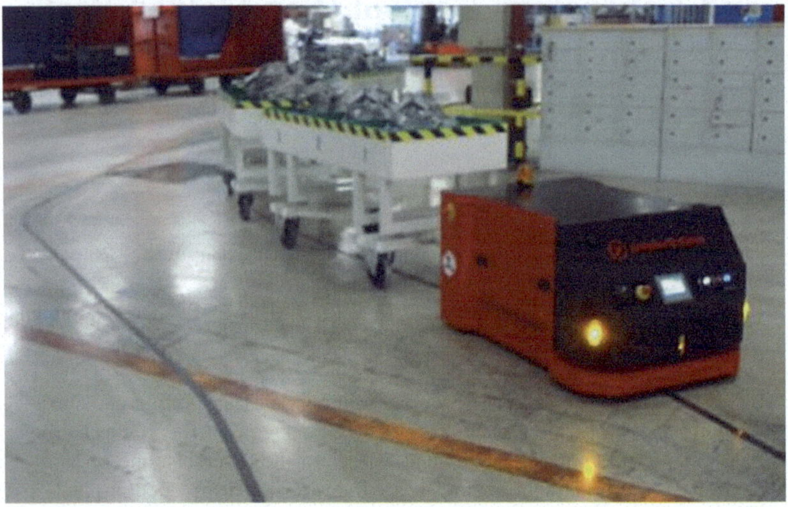

Fig. 1. Automated guided vehicles

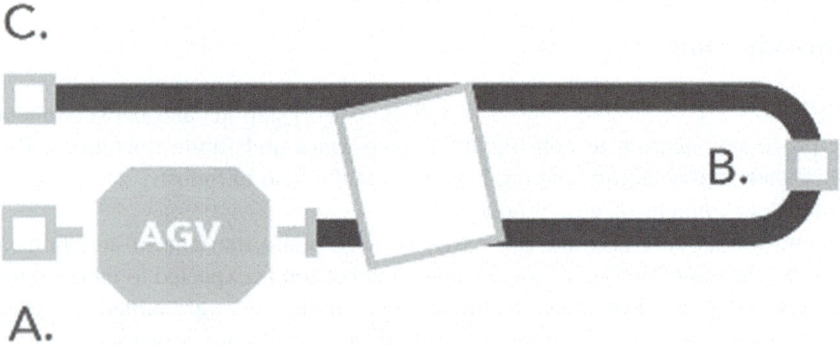

Fig. 2. The route followed by AGV

An AGV (Fig. 1) is an unmanned electric vehicle controlled by pre-programmed software to move parts between different manufacturing and control systems. In the Fig. 2 it can be seen the pre-programmed travel mode of these vehicles. AGVs are based on a magnetically guided sensor Fig. 3 (it is able to detect and report the position of a magnetic field along its axis), beacons, barcodes or predefined laser paths that allow them to move in fixed paths in a controlled space. Lasers and sensors detect obstacles in its path and trigger the vehicle to stop automatically.

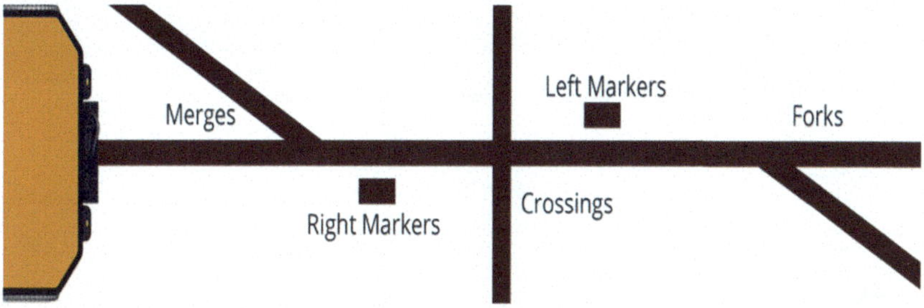

Fig. 3. Magnetic guides

An SDV Fig. 4 is a vehicle that operates without direct route entry or scripts to control steering, acceleration and braking. The secret behind the autonomous navigation of these vehicles is the software. The sensors make the robots perceptive and the algorithms make them intelligent. The software part is the most important component of this system, representing its intelligence. In an industrial environment, an SDV uses laser-based perception and navigation and algorithms to move dynamically between areas of interest. The mode of movement can be seen in the Fig. 5.

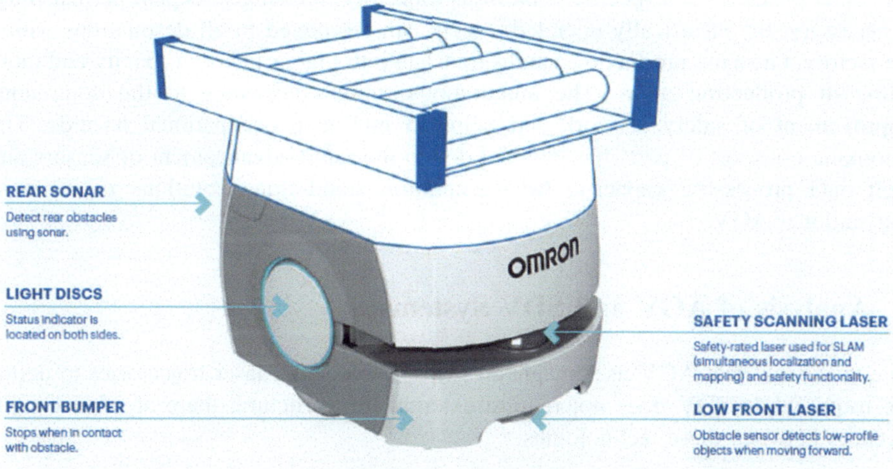

Fig. 4. SDV-self-driving vehicles' (or AMR-autonomous mobile robots)

Fig. 5. The route followed by AGV

The learning ability of vehicle allows them to become more efficient and accurate when meeting new situations. The vehicle navigates its environment with the help of sophisticated cutting edge technology: using sensors and laser scanners, it recognizes vehicles and immobile obstacles, as well as people who get in its way. The control system calculates the approach speed and detects the imminence of a collision. In this case, the robot stops itself or performs an avoidance action. Unlike other systems used, it immediately adapts its route based on environmental information, without having to stop during the process. If the fully autonomous robot detects that it will regularly encounter obstacles at a specific point in its trajectory, it changes its path permanently. If necessary, the electrically operated system can be moved to all destinations within the factory. Growing production figures translate into higher factory capacity and more traffic in production areas. The autonomous robot contributes to the continuous improvement of safety at work and helps to minimize occupational hazards. The automatic transport of parts has evolved due to the rapid advancement of sensors and high data processing capacity. Next-generation stand-alone solutions overshadow conventional AGV.

2 Analysis of AGV and SDV Systems

Due to the fact that AGV uses magnetic strips, barcodes or laser trajectories to define the route, while SDV does not require external infrastructure, there are 5 main differences between these technologies.

2.1 Flexibility and Versatility

An SDV unit can be used in several applications while the number of AGV units depends on the number of applications.

SDV facilitates 5S standards (Fig. 6).

AGV units require a permanent maintenance of the infrastructure.

SDV units, unlike AGVs that if they encounter an obstacle in their way stop until that obstacle is removed, detect, avoid and move dynamically around the obstacles to continue moving to the destination point, human interaction not being necessary. Parameters can be customized to navigate hallways, staff areas and narrow corridors.

Fig. 6.

2.2 Ease of Scaling

An AGV unit may be added to the facility if the scheme meets the infrastructure requirements of the AGV (resources are required for facility planning, infrastructure renovation, maintenance and line training).

Additional SDV units can be operational in a short time because they are controlled by a centralized map that is shared by the other SDV units of the fleet (no infrastructure restoration, facility planning, or additional operator training, no outsourcing required to third party suppliers of additional modification or implementation works) (Fig. 7).

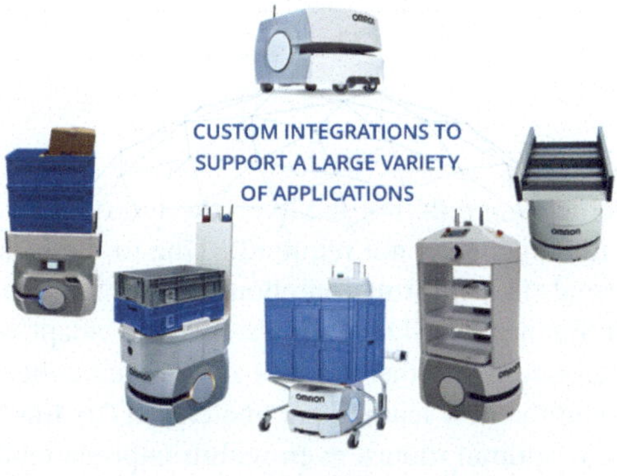

Fig. 7.

2.3 Relocation

Moving an AGV is equivalent to installing an AGV for the first time - significant resources are required. More systems are needed despite long/short seasonal periods.

SDVs can be relocated from one factory to another or to another area of the factory, the configuration time being minimal due to the centrally coordinated fleet manager. The units can be shared between several areas to meet seasonal requirements. Efficiency is increased in facilities or areas, where otherwise it would be a blockage, by redistributing underutilized units from other facilities or areas that face a low need for them. SDVs can become a shared resource, so fewer acquisitions of such units are needed (Fig. 8).

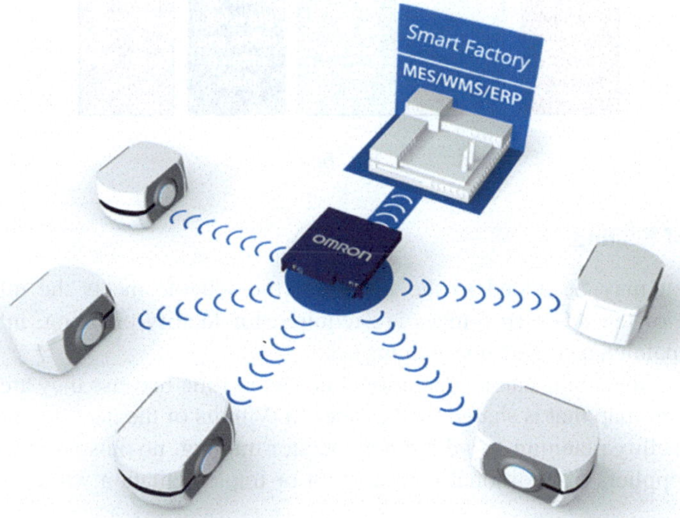

Fig. 8.

2.4 Intelligence

AGVs are not smart vehicles and do not facilitate the industry's vision of Industry 4.0 (due to infrastructure, intelligence is not required). Due to the lack of data acquisition, operators do not have real-time information about the vehicle's performance.

The integrated intelligence of SDVs allows them to adapt to the environmental requirements and to integrate with other solutions. The data collected during operation updates the fleet map with the new learned parameters. SDVs find out which routes are the fastest and follow the optimal routes, even within unpredictable environments. The plant staff can safely interact in collaboration with the SDVs. SDVs provide integrated lights and sounds that resemble those of cars to intuitively indicate behaviors for plant personnel. Due to the fact that they are equipped with collision avoidance sensors in interaction with humans and the possibility of avoiding obstacles encountered, they are considered collaborative robots (cobots).

- Intelligent task assignment. Reduces wasted time and movement by continuously tracking to anticipate which robots will be best positioned for future tasks (Fig. 9).

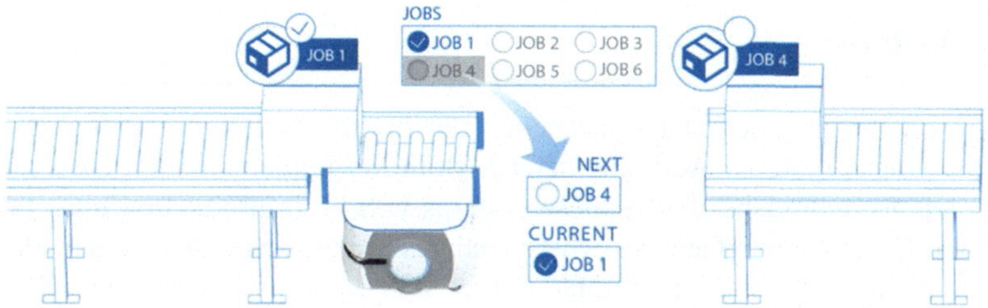

Fig. 9.

- Traffic control. Notify robots of predictable routes, allowing them to recalculate to avoid collisions in the most efficient way (Fig. 10).

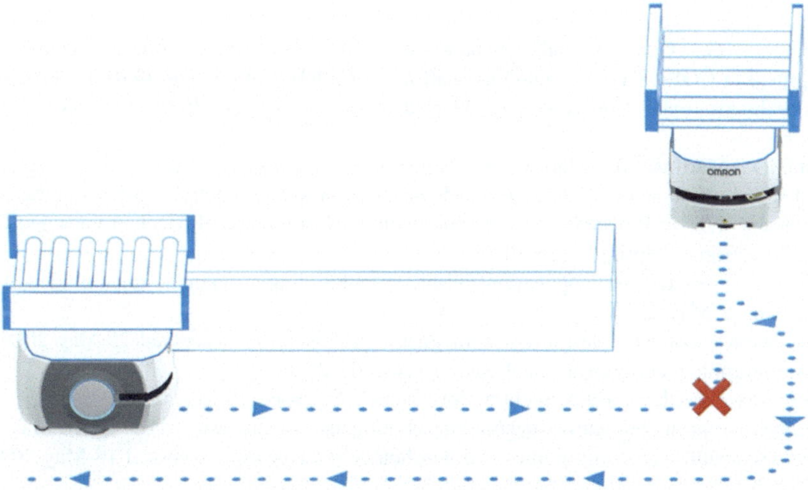

Fig. 10.

2.5 Use

AGVs are designed to do simple tasks, setup and operation are cumbersome, complex, and expensive. They require expertise from engineers or certified personnel. Changes to the system require infrastructure upgrades and additional staff training.

The implementation of the SDV includes the one-time mapping facility with a vehicle and then the setting of areas and points of interest on the map in the fleet administrator. The system is set by the customer and the vehicle or access areas can be updated by the factory staff. Changes in the production line or routes are easily implemented.

3 Conclusions

Depending on the production requirements, the optimal solution is adopted for the plants where these systems are implemented. Both the implementation costs and how long the parts to be transported by these systems will be in production must be taken into account. In the case of factories where changes in production lines are made over a long period of time, it may be preferable to use AGV systems. Taking into account the fact that the trend in the automotive field is to reduce the time to make a new product, SDV systems are increasingly required. The implementation of new parts involves the rearrangement of production spaces as well as a more precise management of stocks and storage spaces. The communication of the SDVs with a centralized system as well as their rapid relocation recommends them for use in these plants.

References

1. Gheorghe, G.: Concept and mechatronics and cyber-mixmechatronics constructions, integrated in COBOT type technology platform for intelligent industry (4.0). In: Proceedings of the International Conference of Mechatronics and Cyber-MixMechatronics. Springer (2019)
2. Costa, D., Martins, M., Martins, S., Teixeira, E., Bastos, A., Cunha, A.R., Varela, L., Machado, J.: Evaluation of different mechanisms of production activity control in the context of industry 4.0. In: Proceedings of the International Conference of Mechatronics and Cyber-MixMechatronics. Springer (2019)
3. https://www.roboteq.com/applications/all-blogs/18-building-a-magnetic-track-guided-agv. Accessed 25 May 2020
4. https://www.roboteq.com/all-products/magnetic-guide-sensors. Accessed 25 May 2020
5. https://ottomotors.com/resources/ebooks. Accessed 22 May 2020
6. http://www.cbimakerspace.com/cbi/5s-for-offices/. Accessed 20 May 2020
7. http://www.omron.com.au/robotics/mobilerobot/business-value.asp. Accessed 18 May 2020
8. http://www.omron.com.au/robotics/mobilerobot/technology.asp. Accessed 18 May 2020

The Dynamic Modeling of Ball and Plate Mechatronic System with Two Simultaneously Degrees of Freedom

Alina Rodica Spânu(✉), Daniel Besnea, and Edgar Moraru

Mechatronic and Precision Engineering Department, University "Politehnica" of Bucharest, Bucharest, Romania
spanu_alina@yahoo.com

Abstract. The paper presents the dynamic mathematical model of the two degrees of freedom movement for the mechatronic system comprising the ball and the plate actuated by servo-electric motors. The main goal is the increasing accuracy by computing and analyzing the results, so that the next step of motion control by using controllers could be done more precisely for tracking an imposed trajectory for instance.

Keywords: Modeling · Non-linear system · Under-actuated mechanism

1 Introduction

During the last years, the control of dynamic systems have become an important challenge due to its spread into the surrounding industrial activities such as industrial robots functioning, transportation, health care and some other fields of research. There are many requirements of accurate positioning for the gripper for instance, where we have to control two or three, may be more electrical motors used for actuating the spatial mechanism.

These systems, implying mechanical systems as well as electrical devices, should be mathematically modeled by using dynamic equations and finally analyzing them from the minimum error point of view. Meantime, in order to increase the accuracy there are some control techniques, using the PD and PID controller, so the resulting movement has to be inside the imposed limits. The main problem of the motion control is to determine the parameters according to the real-time registered position that could be done by using the webcam or some other devices with dedicated software.

The paper [1] presents the research of modeling the servo-vision system made by the ball and plate having two degrees of freedom. The main objective is to analyze the response of the system for each degree of freedom. Additionally they have used two control strategies: the Repetitive Control and the Multiple Resonant Control.

The paper [2] deals with a ball balancer system with two degrees of freedom actuated by two servo-motors simultaneously. The information about the motion is obtained from cameras for surveillance and a PID controller is used for both axes. A better solution could be provided, if a PD controller was included in the system when

G. I. Gheorghe (Ed.): ICOMECYME 2020, LNNS 143, pp. 173–178, 2020.
https://doi.org/10.1007/978-3-030-53973-3_19

the external disturbance with significant values is acting on. The mathematical model of the dynamic system was simplified by decoupling the ball and beam system.

A virtual and remote laboratory for augmented reality was presented in [3] in order to study the non-linear, multi-variable and open-loop unstable system consisting of a ball and plate. The mathematical model was made for the movement with two degrees of freedom for increasing the precision of carrying the ball from a specific position and holding it in this desired final position. They have treated the system as two different subsystems operating simultaneously. Hence, similar but independent controllers can be used for each motion.

The paper [4] presents the motion detailed analysis for a ball that is moving along the beam. The main contribution is the control system, characterized by novel estimate of the domain of attraction to ensure system performance.

The control objective is to stabilize the positions of the ball on the beam while tracking the reference trajectory signal, so the main idea presented in the paper [5] is the control algorithm for the ball that is moving along the beam comprising two degrees of freedom with two transfer functions. Other purpose of this control type is to minimizing the oscillations of the servo angle by using an inner loop and an outer loop. The outer-beam controller controls the position of the ball on the beam and the inner-loop controller controls the angle of the beam. The main advantage is that the two linear controllers are dimensioned separately.

The paper deals with the dynamic study and control of the system that complies with a sphere and a plate, so the ball is rolling free without slipping on the rigid surface. The main goal of this activity is to control the position of the ball by manipulating simultaneously the inclination angles of the plate, so that the imposed trajectory tracking could be achieved with increasing accuracy.

2 The Mathematical Model

In order to propose the mathematical model for the mechatronic system comprising a ball and a plate moving around two directions simultaneously, we have analyzed the system presented in Fig. 1.

For the mathematical model we will take into account the following considerations:

- the ball is a symmetric and homogenous sphere;
- the ball should be in contact with the plate during the movement;
- the ball is moving over the plate without slipping;
- the plate is rigid and homogenous.

In Fig. 1 the notations are the following: 1 – servo-electric motor for the movement around OX axis; 2 – servo-electric motor for the movement around OY axis; 3 – horizontal plate; 4 – sphere; XOYZ – the fixed Cartesian system. The information regarding the position of the ball along the plate could be obtained from a webcam, so that we may know the coordinate of the contact point between the ball and the plate measured in the fixed Cartesian system.

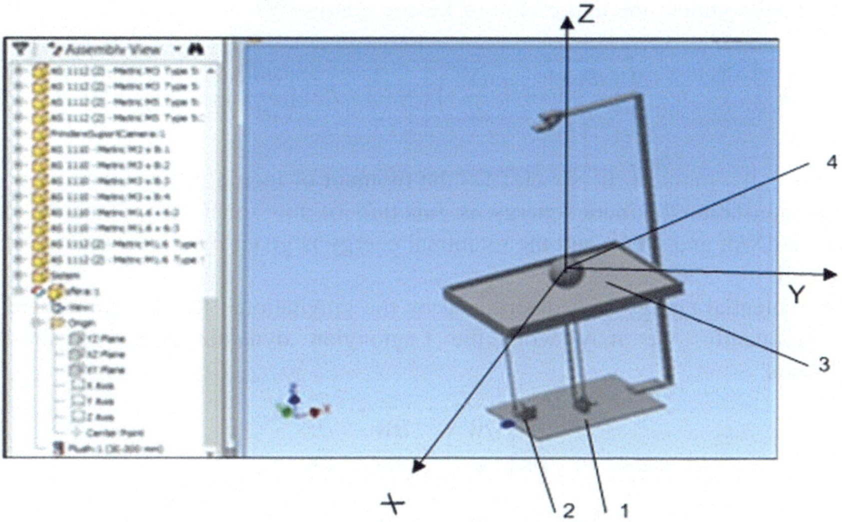

Fig. 1. The three-dimensional model of the system.

Using the notations represented in Fig. 1, we may write the condition that the contact point of the sphere has the coordinates X, Y measured in the fixed Cartesian system before the beginning of the motion. After the movement has started, the contact point will have the coordinates x_1, y_1 as function of coordinates in the fixed system XOYZ. If we consider θ the rotational angle around OX and β the rotational angle around OY we may assume:

– for the first movement:

$$\begin{cases} x_1 = X \cdot \cos(\theta) \\ y_1 = Y \cdot \sin(\theta) \end{cases} \tag{1}$$

– for the second movement simultaneously with the first movement:

$$\begin{cases} x_1 = X_0 \cdot \cos(\theta) \cdot \sin(\beta) \\ y_1 = Y_0 \cdot \sin(\theta) \cdot \cos(\beta) \end{cases} \tag{2}$$

We may conclude the ball position is a function of both movements around OX axis and around OY axis.

The mathematical model of the sphere motion could be written using the Lagrangian formalism with θ and β as generalized coordinates, as well as their derivatives for the rotational speed and acceleration.

We may compute the expression of kinetic energy W:

$$W = \frac{1}{2} \cdot m \cdot \left[\left(\frac{dX}{dt} \right)^2 + \left(\frac{dY^2}{dt} \right) \right] + \frac{1}{2} \cdot J \cdot \left(\frac{d\theta^2}{dt} \right) + J \left(\frac{d\beta^2}{dt} \right) \cdot \frac{1}{2} \tag{3}$$

where: m – the mass of the sphere; J – the moment of inertia of the sphere;

The translational kinetic energy is function of the sphere mass and its linear velocities dX/dt and dY/dt and the rotational energy is given by the moment of inertia of the ball.

The potential energy U is determined by the gravitational force of the sphere.

Consequently, we may write the Lagrangian dynamic equations for both movements:

$$\frac{\partial}{\partial t}\left(\frac{\partial W}{\partial \dot{\theta}}\right) - \frac{\partial W}{\partial \theta} = \frac{\partial U}{\partial \theta} \tag{4}$$

$$\frac{\partial}{\partial t}\left(\frac{\partial W}{\partial \dot{\beta}}\right) - \frac{\partial W}{\partial \beta} = \frac{\partial U}{\partial \beta} \tag{5}$$

Taking into account all the mathematical equations written above, we have computed the following system of four differential equations:

$$\begin{aligned} \frac{d^2\theta}{dt^2} = &\frac{1}{[m \cdot (X^2 \cdot sin^2(\theta) \cdot sin^2(\beta) + Y^2 \cdot cos^2(\theta) \cdot cos^2(\beta))]} \\ &\cdot [m \cdot \dot{\theta} \cdot (Y^2 \cdot \sin(2\theta) \cdot \dot{\theta} \cdot cos^2(\beta) + Y^2 \cdot \sin(2\beta) \cdot cos^2(\theta) \cdot \dot{\beta} \\ &- X^2 \cdot \sin(2\theta) \cdot \dot{\theta} \cdot sin^2(\beta) - X^2 \cdot \sin(2\beta) \cdot sin^2(\theta) \cdot \dot{\beta}] - m \cdot g \cdot \sin(\theta) \\ &+ m/2 \cdot (X^2 \cdot \sin(2\theta) \cdot \dot{\theta} \cdot sin^2(\beta) - X^2 \cdot \sin(2\theta) \cdot cos^2(\beta) \cdot \dot{\beta}^2 \\ &- Y^2 \cdot \sin(2\theta) \cdot cos^2(\beta) \cdot \dot{\theta}^2 + Y^2 \cdot \sin(2\theta) \cdot sin^2(\beta) \cdot \dot{\beta}^2) - J\dot{\theta} \cdot] \end{aligned} \tag{6}$$

$$\frac{d\theta}{dt} = \dot{\theta} \tag{7}$$

$$\begin{aligned} \frac{d^2\beta}{dt^2} = &\frac{1}{[m \cdot (X^2 \cdot cos^2(\theta) \cdot cos^2(\beta) + Y^2 \cdot sin^2(\theta) \cdot sin^2(\beta))]} \\ &\cdot [m \cdot \dot{\beta} \cdot (X^2 \cdot \sin(2\beta) \cdot \dot{\beta} \cdot cos^2(\theta) + X^2 \cdot \sin(2\theta) \cdot cos^2(\beta) \cdot \dot{\theta} \\ &- Y^2 \cdot \sin(2\theta) \cdot \dot{\theta} \cdot sin^2(\beta) - Y^2 \cdot \sin(2\beta) \cdot sin^2(\theta) \cdot \dot{\beta}] + m \cdot g \cdot \sin(\beta) \\ &+ \tfrac{m}{2} \cdot (\sin(2\beta) \cdot \dot{\theta}^2 \cdot (X^2 \cdot sin^2(\theta) - Y^2 \cdot cos^2(\theta)) + \sin(2\beta) \cdot \dot{\beta}^2 \\ &\cdot (X^2 \cdot cos^2(\theta) + Y^2 \cdot sin^2(\theta)) - J\dot{\beta})] \end{aligned} \tag{8}$$

$$\frac{d\beta}{dt} = \dot{\beta} \tag{9}$$

In the above written system we have considered that X and Y are the coordinates of contact point between the sphere and plate measured in the fixed Cartesian system. The computed results are: θ – the angular value for the movement around X axis; β – the angular value for the movement around Y axis; $\dot{\theta}$ - the rotational speed around X axis; $\dot{\beta}$ - the rotational speed around Y axis.

3 The Experimental Set-Up and Modeling Results

The experimental set-up is presented in Fig. 2 and Fig. 3, so we may observe the two servo-electric motors mounted on the fixed base at the bottom, the arms with three spherical joint, the webcam at the top of the system, the ball and the plate.

Fig. 2. The experimental set-up: the electric motors

Fig. 3. The plate and the webcam.

We have considered the following numerical values for the mechatronic system: the sphere with the mass m = 0.11 [Kg] and the moment of inertia $J = 9.9 \cdot 10^{-6}$ [Kgm^2]; the coordinates of contact points X = 10 mm and Y = 10 mm. The initial values: $\theta = 10$ [deg] and $\beta = 5$ [deg].

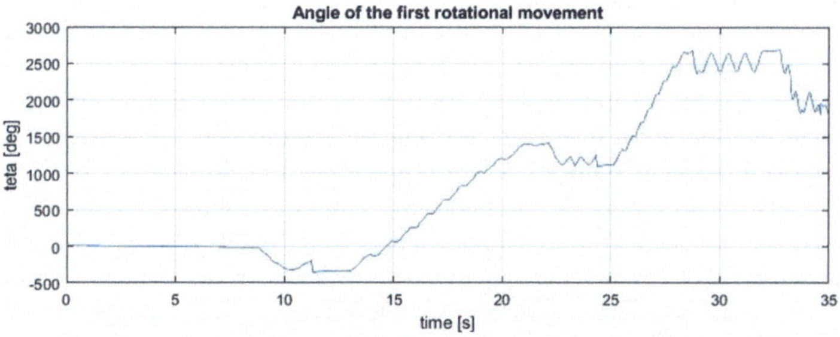

Fig. 4. The variation of angle theta as time function.

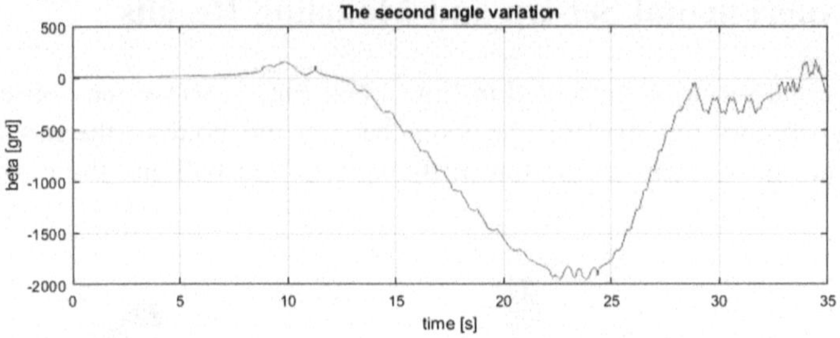

Fig. 5. The variation of the second angle as time function.

As we may infer from the Fig. 4 and Fig. 5 the angle variations are following few steps during an increasing period, at first there are 2.5 [s], later 5 [s] and finally 10 [s]. During each step, there are some oscillations until the movement is stabilized, so we may improve this characteristic by using the PD or PID controller. Meantime, by analyzing the shape of variations presented in Fig. 4 and Fig. 5, we may admit the time for passing through the phases of motion is very short.

4 Conclusions

The paper present the mathematical model of the mechatronic system complying the ball and the plate actuated simultaneously by two servo-electric motors. Starting from the Lagrangian formalism we have written the mathematical model of the dynamics functioning, so that the positioning accuracy could be improved. If we know the position of the ball along the plane of the plate, we may determine the angles for each servo-electric motor, in order to keep the sphere tracking the imposed trajectory.

References

1. Castro, R., Flores, J., Salton, A., Pereira, L.F.A.: A comparative analysis of repetitive and resonant controllers to a servo-vision ball and plate system. In: 19th Congress International Federation of Automatic Control, pp. 1120–1125, Cape-Town (2014)
2. Arun, K.P., Geetha, M., Chandran, K.R., Karthik, P.: Composite disturbance rejection control for ball balancer system. In: International Conference on Robotics and Smart Manufacturing (RoSMa 2018), Procedia Computer Science, pp. 124–133 (2018)
3. Fabregas, E., Dormido-Canto, S., Dormido, S.: Virtual and remote laboratory with the ball and plate system. Int. Fed. Autom. Control, PapersOnLine **50–1**, 9132–9137 (2017)
4. Kelly, R., Sandoval, J., Santibanez, V.: A novel estimate of the domain of attraction of an IDA-PBC of a ball and beam system. In: Proceedings of the 18th World Congress, the International Federation of Automatic Control, Milano, Italy, pp. 8463–8467 (2011)
5. Mehedi, I.M., Al-Saggaf, U., Mansouri, R., Bettayeb, M.: Two degrees of freedom fractional controller design: application to the ball and beam system. Measurement **135**, 13–22 (2019)

Wideband Bandpass Filter Design Based on RF-MEMS Technology

Syed M. Sifat[1], Raj Savaj[2], Ion Stiharu[2(✉)], and Ahmed Kishk[1]

[1] ECE Department, Concordia University, Montreal, QC, Canada
s_sifat@encs.concordia.ca
[2] MIAE Department, Concordia University, Montreal, QC, Canada
istih@encs.concordia.ca

Abstract. In this paper, we shall present the design steps and analyze the performance of a wideband bandpass filter based on RF MEMS technology. MEMS technology enables very accurate features, which enables the repeatability of the filter. The filter is configured as a parallel edge coupled five-pole microstrip wideband bandpass filter. The concept consists of using precisely sized cantilever beams to excite the filters, which that will act as a switch to feed the filter. The bimetallic switch uses a cantilever beam to perform the deflection based on temperature rise. The filter part is designed using CST Microwave Studio (Frequency Domain Solver), and the cantilever beam is designed using AUTOCAD.

Keywords: RF-MEMS · Cantilever beam · Bandpass filter · Bimetallic switch

1 Introduction

Microelectromechanical systems (MEMS) have gained its prominence and becoming very attractive in displaying superior performance in building compact devices and components. Microelectromechanical systems (MEMS) are miniature devices combining electrical and mechanical components and fabricated using integrated circuits (IC) batch-processing techniques [1, 2]. In terms of fabrication, the surface micromachining and bulk micromachining are the most commonly used techniques. Surface micromachining consists of the deposition and lithographic patterning of various thin films, usually on silicon substrates as illustrated in Fig. 1. The deposition can be performed by adding a “sacrificial film” underneath the one which needs to be released and removed in the last steps of the process by selective etching techniques. The variety of structural materials is enormous, including many metals (Au, Al, etc.), ceramics (SiO_2 and Si_3N_4), and plastics such as photoresist, polymethyl methacrylate (PMMA), and others [3].

On the other hand, for bulk micromachining, a silicon substrate wafer is selectively etched to create the structures. The process includes the steps of wet chemical etching, reactive-ion etching (RIE), or both to form the released or fixed microstructures. With wet etching, the resulting structures depend on the directionality of the etching, which is a function of the crystallinity of the substrate and the etching chemistry [4]. Figure 2 illustrates the bulk micromachining technique with all the required fabrication steps.

G. I. Gheorghe (Ed.): ICOMECYME 2020, LNNS 143, pp. 179–187, 2020.
https://doi.org/10.1007/978-3-030-53973-3_20

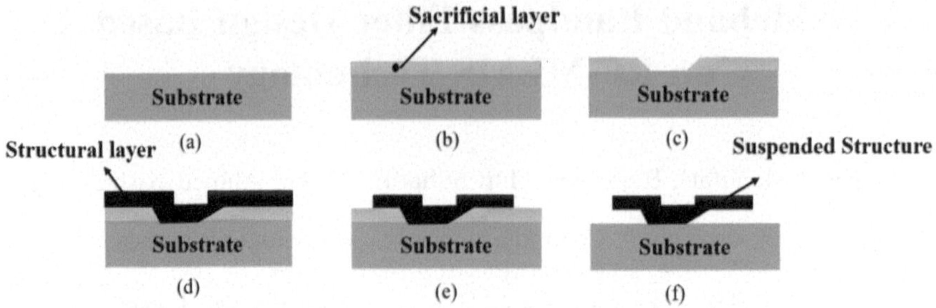

Fig. 1. Typical steps in a surface micromachining process, (a) silicon substrate, (b) deposition of sacrificial layer, (c) etching sacrificial layer, (d) deposition of the structural layer, (e) shape definition by patterning and etching, (f) release of suspended structure.

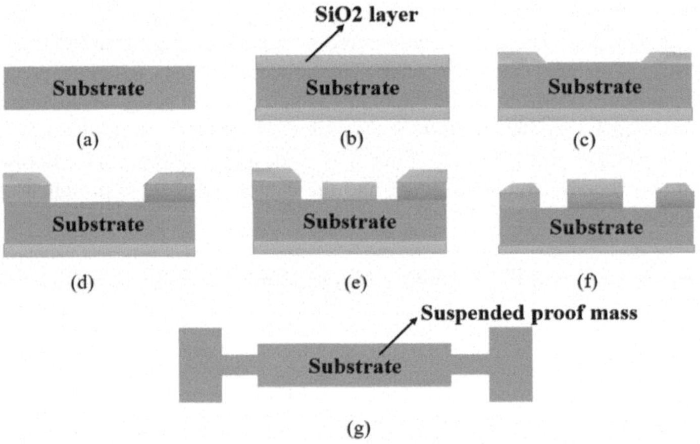

Fig. 2. Typical steps in a bulk micromachining process (dry etching): (a) silicon substrate preparation, (b) deposition of SiO_2 layer, (c) etching of SiO_2 layer (hard mask, when necessary), (d) substrate etching, (e) deposition of SiO_2 layer for a selective area, (f) substrate etching, (g) creation of suspended structure. The figure shows from a) to f) only one side patterning. The bottom could be processed simultaneous with the top or one side at the time to achieve the geometry as in (g).

Microelectromechanical Systems (MEMS) have been developed and improved since the early 1970's for various application purposes such as pressure and temperature measurement, as accelerometers, gas sensors, radiation sensors and others. MEMS switches for low-frequency applications have been deployed in the early 1980's. Later, in the beginnings of 1990's, MEMS switches were developed for microwave applications [5]. MEMS devices are now used in large fields of applications including automotive, biomedical, consumer, aerospace and military applications. In wireless communication devices, RF MEMS devices such as phase shifters, filters, on-chip antennas, and a tunable matching network for wideband radios are generally used [6]. The RF MEMS switches are ideal for low power reconfigurable networks and subsystems, given the low insertion loss, and high Q-factor. RF switches are easily integrated with antennas, filters, and low-loss matching networks [7–11]. The current CMOS substrate technologies are not suited for integrating antennas; the silicon substrate has a low resistivity – in the range of 10 Ω-cm - which is beneficial for the IC design, while it prevents latch-up currents. However, once the antenna radiates, most of the energy will pass by the low resistivity substrate, and will be absorbed/dissipated, rather than radiating to the free space [12]. In addition, a high amount of surface waves will travel in the substrate that might increase the losses.

Recent developments in micromachining techniques have resulted in novel high performance, low-loss filters for microwave and millimeter-wave applications [13–16]. There are different types of micro-machined filters. One particular example is based on the concept of suspending the microstrip or strip-line on thin dielectric membranes (typically 1.5 μm thick) to eliminate dielectric loss and dispersion problems, resulting in a pure TEM (Transverse Electro Magnetic) mode of propagation and conductor-loss limited performance [14]. In addition, a membrane fabrication and cavity formation, a three-layer structure of SiO_2-Si_3N_4-SiO_2 is deposited on a high-resistivity silicon substrate using thermal oxidation and high-temperature chemical vapor deposition [15]. Park et al. [16] developed a tunable mm-wave bandpass filter using cantilever-based Au varactors, which produces a tuning range of 0.8 GHz (2.5%) at 32 GHz. The filter was built on a glass substrate, and it was based on a lumped element design.

In this work, we will present the design steps and perform the performance analysis of a parallel edge-coupled bandpass filters in RF-MEMS technology, which can work for UWB (Ultra-Wide Band) applications. For the simulation of the bandpass filter, we have used CST MICROWAVE STUDIO® of Dassault Systems. The bandpass filter can be excited with the help of a bimetallic switch that uses a micro-cantilever beam to perform the deflection based on the rise in temperature. The cantilever beam is fixed at one end and free at the other end. The organization of the paper is as follows: Sect. 2 describes a detailed description of the proposed filter. This will follow by the cantilever beam design and analysis in Sect. 3. Finally, Sect. 4 provides a summary of the contributions of the paper.

2 Parallel Edge-Coupled Five-Pole Bandpass Filter

Figure 3 illustrates the schematic of the proposed parallel end-coupled bandpass filter. The bandpass filter is designed using CST Microwave Studios, which uses a Finite Element Method (FEM) solver. The filter is printed on a PCB material having a dielectric constant, ε_r of 10.2, with a thickness of h = 0.15 mm. In general, the parallel-coupled microstrip bandpass filters use half-wavelength line resonators. All the elements are positioned so that adjacent resonators are parallel to each other along half of their length. This parallel arrangement gives relatively large coupling for a given spacing between resonators. Thus, filter structure is particularly convenient for wide bandwidth as compared to the structure with end-coupled microstrip filters. The proposed bandpass filter was designed using reference [17], pp. 127–129. It is worth mentioning, the sole reason behind choosing five elements is for obtaining a wide bandwidth. Chebyshev transformer is one of the most convenient matching transformer that provides an excellent matching bandwidth. The parallel coupled bandpass filter is planar, compact, easy to fabricate, and can be considered for the full-duplex communication in the wireless transmitter, and receivers end. It is also possible to integrate this kind of filter with some other existing guiding structure like Ridge gap waveguide technology where the propagation medium is air instead of a dielectric medium [18, 19].

Figure 4 illustrates the S-parameters of the proposed filter. The reflection coefficient S_{11} of the bandpass filter is covering 15.7% impedance bandwidth from 34.2–40.1 GHz. The insertion loss S_{21} is around −0.6 dB, which is considered very suitable in terms of transmitting power from one end to the other end. At the frequency above and below the resonance the bandpass filter would not allow any signal to pass, which is clearly visible from the S-parameter as above mentioned, is illustrated in Fig. 4. It is important to point out that a waveguide port represents the ideal scenario for solving an electromagnetic problem; however, the accurate way to solve it is by using a co-axial connector. Waveguide ports represent a special kind of boundary condition of the calculation domain, enabling the stimulation as well as the absorption of energy.

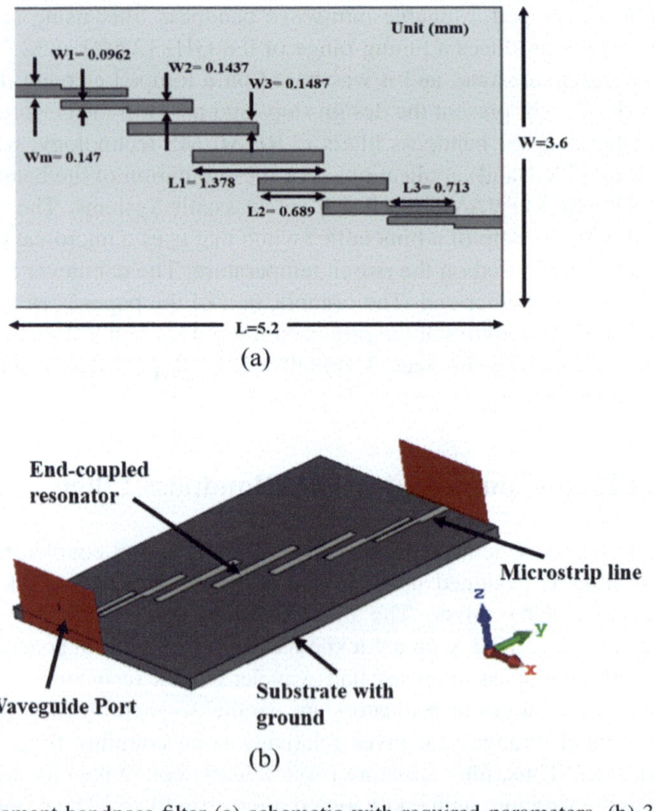

Fig. 3. Five-element bandpass filter (a) schematic with required parameters, (b) 3D geometry.

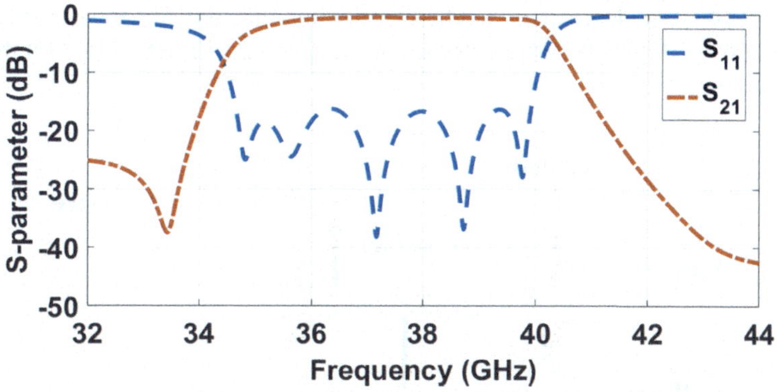

Fig. 4. Scattering (S) parameters of the parallel end-coupled bandpass filter with five elements.

Figure 5 illustrates the Electric field distribution of the bandpass filter at four different frequencies. It is evident from the figure that at 34 and 42 GHz, which are on the top-bands, the filter is not allowing any signal to pass from Port 1 to Port 2. Within the bandpass frequency range, the coupling is strong between the input and the outputs through the parallel-coupled resonators.

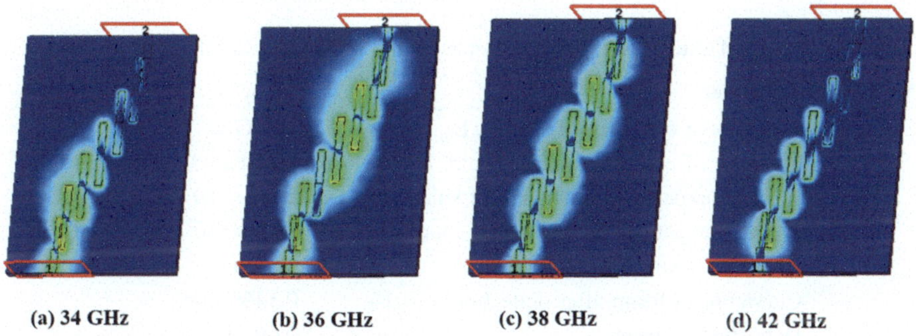

Fig. 5. Electric (E) field distribution at various frequencies of the bandpass filter.

3 Cantilever Beam Design - Bimetallic Switch

MEMS bimetallic switch is used given its ability to deflect when the temperature changes, which occurs without any mechanical force. The function of this switch is to control the openness and closeness of the electrical circuit [20, 21]. Figure 6 shows a bimetallic switch model that is normally used in electrical circuits. MEMS bimetallic switch is one cantilever beam that is fixed at one end and free at the other end. MEMS bimetallic switch is made of two metals joined in a true metallurgical bond enabled by the physical chemical deposition of two dislike metals. When temperature changes, the beam which is the switch, bends due to the different thermal expansion coefficients of both materials. The switch will bend towards the material having a lower thermal expansion coefficient. Here it is worth mentioning that the lower side (blue color) material might be doped polysilicon while the upper side (green color) might be any metallic material.

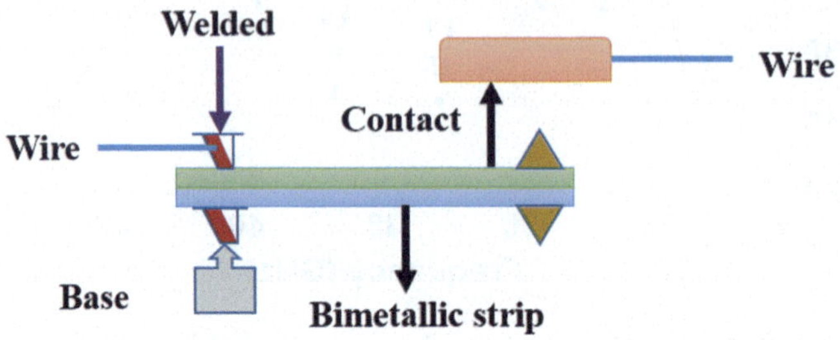

Fig. 6. Bimetallic switch with cantilever beam in electrical circuit.

The width, length, and thickness, of the cantilever beam is W = 0.147 mm, L = 1.1 mm, h = 0.01 mm respectively. The polysilicon and the metallic layers are both kept at the thickness of the filter $\boldsymbol{h}_{Si} = \boldsymbol{h}_{Al} = 0.005$ mm. The selected sizes will enable a reliable operation of the system. The design parameters of the designed bimetallic switch are illustrated in Table 1.

Table 1. Design parameters of the bimetallic switch

Parameter	Value
Modulus of elasticity of silicon, E_{Si}	176 GPa
Modulus of elasticity of aluminum, E_{Al}	80 GPa
Thermal expansion coefficient of silicon, α_{Si}	$2.76 \times 10^{-6}\ k^{-1}$
Thermal expansion coefficient of aluminum, α_{Al}	$23.1 \times 10^{-6}\ k^{-1}$
Radius of beam after deflection, r	6.8 mm
Curvature of beam after deflection, k = 1/r	0.1466 mm^{-1}
Deflection of beam, y	0.09 mm

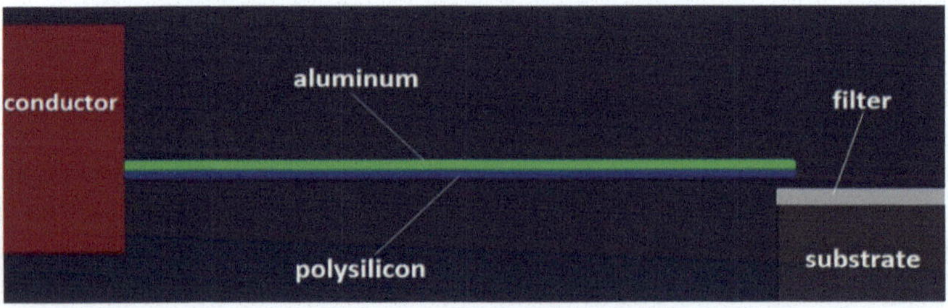

Fig. 7. AutoCAD 2D geometry of bimetallic switch with filter.

The AutoCAD model of the bimetallic switch model is illustrated in Fig. 7 with the cantilever beam. Assuming 50 °C temperature rise for calculation, $\Delta T = 50$ °C, the theoretical curvature equation is calculated using [21],

$$k = \frac{6E_sE_{Al}(h_{si} + h_{Al})h_{si}h_{Al}\varepsilon}{E_{Si^2}h_{Si^4} + 4E_{Si}E_{Al}h_{Si^3}h_{Al} + 6E_{Si}E_{Al}h_{Si^2}h_{Al^2} + 4E_{Si}E_{Al}h_{Al^3}h_{Si} + E_{Al^2}h_{Al^4}} \tag{1}$$

Where, $\varepsilon = (\alpha_{Al} - \alpha_{Si})\Delta T$

Curvature, k is related to deflection, y by the following equation,

$$k = \frac{2\sin(\tan^{-1} y/_l)}{\sqrt{l^2 + y^2}} \tag{2}$$

Hencc, calculations show that the free end of this bimetallic switch deflects by 0.09 mm for a 50 °C temperature rise. It should be mention that these values are calculated using MATLAB coding. As Si has a lower thermal expansion coefficient than aluminum, so the beam as scaled will bend 0.09 mm downside. Therefore, the beam will be positioned to touch the filter that makes the circuit close and current flows through the filter.

One end of the beam is joined with the conductor, as illustrated in Fig. 7. When voltage is supplied, it will come through this conductive material. Material with a high resistivity requires small amount of current, and enough amount of heat is produced due to high resistivity, which will increase temperature. The more the temperature rise, the more beam bends. The heat produced by the current flow is $q = I^2$ Rt (Joule's equation of electric heating), where I = Current (A), R = Resistance (Ω), and t = Time (S). Conductive heat transfer per unit length is $q = kA\Delta T$ (according to Fourier's law) where K = Overall conductive heat transfer coefficient (W/mK), A = Cross-sectional area perpendicular to the current flow (m^2), ΔT = Change in temperature (k).

The deflection of the cantilever beam is calculated using MATLAB coding which is highlighted in Fig. 8.

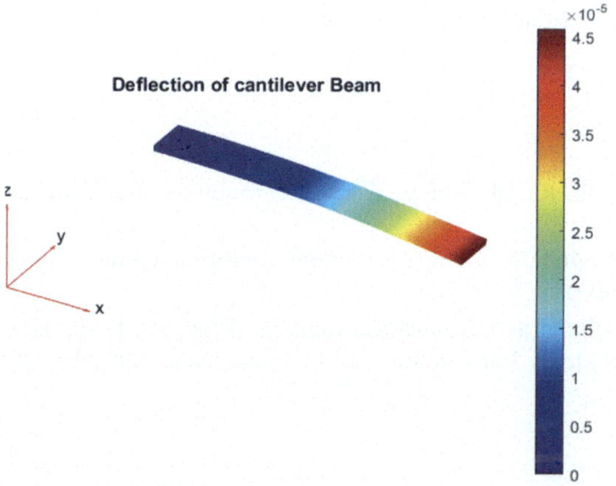

Fig. 8. Deflection of the cantilever beam at 50 °C.

For 50 °C increment in temperature the deflection is 0.09 mm. The more temperature increases the more deflection will occur. For example, for 70 °C temperature increment the deflection will be 0.12 mm which is more than the one encountered at 50 °C as shown in Fig. 9.

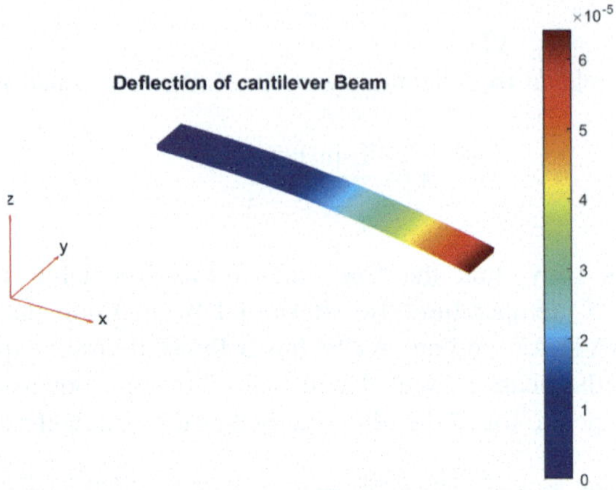

Fig. 9. Deflection of the cantilever beam at 70 °C.

4 Conclusion

In this work, a wideband bandpass filter has been designed covering 15.7% impedance bandwidth from 34.2–40.1 GHz. The bandpass filter can be excited with the help of a bimetallic switch that uses a cantilever beam to perform the deflection based on the rise in temperature. This approach can be feasible for MEMS integrated circuit elements. It is possible to integrate cantilever beam switching with some other existing passive elements such as antennas, couplers, circulators, and power dividers from integrated circuits. Bandpass filters are widely used in wireless transmitter and receiver ends, and for ultrawideband full-duplex commutation.

References

1. Bryzek, J., Peterson, K., McCulley, W.: Micromachines on the March. IEEE Spectr. **31**(5), 20–31 (1994)
2. Brown, E.R.: RF-MEMS switches for reconfigurable integrated circuits. IEEE Trans MTT **46**, 1868–1880 (1998)
3. Peterson, K.E.: Silicon as a mechanical material. Proc. IEEE **70**, 420–457 (1982)
4. Gad-el-Hak, M.: Mems Introduction and Fundamentals, 2nd edn. CRC Press, Boca Raton (2006)

5. Larson, L.E., Hackett, R.H., Melendes, M.A., Lohr, R.F.: Micromachined microwave actuator (MIMAC) technology-a new tuning approach for microwave integrated circuits. In: IEEE Microwave and Millimeter-Wave Monolithic Circuits Symposium., Boston, MA, USA, pp. 27–30 (1991)
6. Rebeiz, G.M.: RF MEMS Theory, Design, and Technology. Wiley, New Jersey (2014)
7. Brown, A.R., Rebeiz, G.M.: A high-performance integrated K-band diplexer. IEEE Trans. Microw. Theory Tech. **47**(8), 1477–1481 (1999)
8. Ruby, R.C., et al.: High-Q FBAR filters in a wafer-level chip scale package. In: IEEE International Solid-State Circuits Conference, San Francisco, CA, USA, pp. 184–458 (2002)
9. Bannon, F.D., Clark, J.R., Nguyen, C.T.: High-Q HF microelectromechanical filters. IEEE J. Solid-State Circuits **35**(4), 512–526 (2000)
10. Kim, M., Hacker, J.B., Mihailovich, R.E., DeNatale, J.F.: A monolithic MEMS switched dual-path power amplifier. IEEE Microw. Wirel. Compon. Lett. **11**(7), 285–286 (2001)
11. Erdil, E., Topalli, K., Unlu, M., Civi, O.A., Akin, T.: Frequency tunable microstrip patch antenna using RF MEMS technology. IEEE Trans. Antennas Propag. **55**(4), 1193–1196 (2007)
12. Hammad, M.C., Atif, S.: The last barrier. IEEE Microw. Mag. **10**, 79–91 (2013)
13. Rebeiz, G.M., Katehi, L.P., Weller, T.M., Chi, C.Y., Robertson, S.V.: Micro machined membrane filters for microwave and millimeter-wave applications. Int. J. Microw. Millim.-Wave CAE **7**, 149–166 (1997)
14. Weller, T.M., Katehi, L.P.: Miniature stub and filter designs using the micro shield transmission line. In: IEEE MTT-S, Digest, pp. 675–678 (1995)
15. Blondy, P., Brown, A.R., Cros, D., Rebeiz, G.M.: Low loss micro machined filters for millimeter-wave communication systems. IEEE Trans. MTT **46**, 2283–2288 (1998)
16. Park, J.-H., Kim, H.-T., Kwon, Y., Kim, Y.K.: Tunable millimeter wave filters using coplanar waveguide an micromachined variable capacitors. Micro-Eng. Micromech. **11**, 706–712 (2001)
17. Hong, J.-S., Lancaster, M.J.: Microstrip Filters for RF/Microwave Applications. Wiley, Hoboken (2001)
18. Sorkherizi, M.S., Kishk, A.A.: Self-packaged, low-loss, planar bandpass filters for millimeter-wave application based on printed gap waveguide technology. IEEE Trans. Compon. Packag. Manuf. Technol. **7**(9), 1419–1431 (2017)
19. Sifat, S.M., Ali, M.M.M., Shams, S.I., Sebak, A.: High gain bow-tie slot antenna array loaded with grooves based on printed ridge gap waveguide technology. IEEE Access **7**, 36177–36185 (2019)
20. Al-Dahleh, R., Mansour, R.R.: A novel warped-beam design that enhances RF performance of capacitive MEMS switches. In: IEEE/MTT-S International Microwave Symposium, Honolulu, HI, pp. 1813–1816 (2007)
21. Clyne, T.: Residual stresses in surface coatings and their effects on interfacial debonding. Key Engineering Materials (Switzerland), vol. 116 (1996)

Energy Efficient Network Manufacturing System Using Controlled Elitist Non-dominated Sorting Genetic Algorithm

Veera Babu Ramakurthi[1], V. K. Manupati[1], Leonilde Varela[2], and José Machado[3(✉)]

[1] Department of Mechanical Engineering, National Institute of Technology Warangal, Hanamkonda, India
rvbabu.nitw@gmail.com, manupati.vijay@nitw.ac.in
[2] Algoritmi Research Centre, University of Minho, Braga, Portugal
leonilde@dps.uminho.pt
[3] MEtRICs Research Centre, University of Minho, Braga, Portugal
jmachado@dem.uminho.pt

Abstract. Recent manufacturing systems did not just confine to optimal utilization of resources due to the global stance on strict environmental regimes. Collaborative effort to achieve sustainable practices in the decentralized manufacturing environment is a new complex problem. In this paper, with a networked manufacturing system we try to achieve both traditional as well as sustainable parameters by optimizing the performances such as makespan, machine utilization, and energy consumption. Thereafter, we formulate the problem as a mixed-integer non-linear programming (MINLP) model. To handle this NP-hard problem and to find the optimal solutions a Controlled elitist non-dominated sorting genetic algorithm (CE-NSGA-II) has been adopted. Finally, the results are analyzed with different scenarios to prove the proposed approach validation.

Keywords: Networked manufacturing system · Sustainability · Genetic algorithm · Optimization

1 Introduction

Since most resources are non-renewable, globalization and rapid development in developing countries have led to increased consumption in energy resources. The industrial manufacturing sectors are consuming almost half of the energy delivered by the world according to Environmental Impact Assessment (EIA). In addition, countries like China, India, and Brazil, there is a global demand for a large variety of goods due to their relatively higher population growth and development in overall living standards. However, to fulfill their needs, the resources required are quite scarce. Hence, the efficient and sustainable utilization of resources has to be adopted, especially in the manufacturing sector (Haapala et al. 2013). Sustainable development has been defined as development to satisfy the needs of the present without affecting the ability of future

G. I. Gheorghe (Ed.): ICOMECYME 2020, LNNS 143, pp. 188–206, 2020.
https://doi.org/10.1007/978-3-030-53973-3_21

generations to meet their own needs by taking into account economic, social, and environmental dimensions (Jovane et al. 2008).

In addition, markets should sustain the benefits of enormous market competition owing to the fact that increased intricacy and functionality of vastly demanded products. The overall economic contest has certain advantages such as reduced cycle-time in the manufacturing system, ultimate data information, the standard flow of knowledge, etc. In order to achieve the mentioned requirements, current conventional manufacturing scenarios have to be renovated so that globally growing customer's demand can be satisfied. In this study, a Networked manufacturing (NM) environment has considered to gain the advantage that lacks in the conventional manufacturing system. In this research, a multi-objective network-based manufacturing model of customers, enterprise users, and a cloud of enterprises are developed to optimize makespan, machine utilization, and energy consumption while disclosing a product.

Several new-generation manufacturing systems emerged in recent times which can capable enough to adopt to changes in environmental market conditions, especially when numerous turbulent variations are evolved in the market demand (Peklenik and Jerele 1992). However, an enterprise can attain improved quality and cost-efficient manufacturing schedule by boosting the manufacturing system's reconfigurability and flexibility through introducing a distributed paradigm (Veeramani 1997; Wilde and Briscoe 2011). These specified needs and their functionalities can be upheld by applying networked manufacturing approach. The definition, functions, advantages, applications and its limitations are detailed in different studies (Varela et al. 2018).

Networked manufacturing situations quite different from the monolithic approach in job scheduling criterion. Hence, the idea of a conventional job scheduling concept is expanded and reorganized. In networked-based manufacturing, various jobs consist of competition issues in between them so individual optimal results are generated which is different from the traditional case of manufacturing. However, the characteristics of real-world production atmosphere are stochastic and this system is perfectly distinguished by taking many optimizing objective functions concurrently. As a result, efficient algorithms are necessary for a wide-range thus optimal solutions can be achieved in lesser computational run time.

Ausaf *et al.* (2015) presented a priority-based optimization algorithm (PBOA) that utilizes a mixture of operation engagement prioritization along with a structure of dispatching rules to help in efficiently generating a schedule for MOIPPS. Considering the intricacies of practical IPP (Integrated Production Planning) like wide-reaching decision variables, several objective functions, and uncertainty in interval-valued parameters. Lin *et al.* (2016) introduced a novel "order-set" concept to be used with modified interval-MOEA pertaining to steel-making continuous casting-hot rolling process and it was implemented in an iron-steel company of China after testing with their routine production data. Luo *et al.* (2017) proposed a multi-objective genetic algorithm (MOGA) based on the immune principle associated with density mechanism and external archive to establish a much better IPPS with maintained diversification of population, avoided premature condition and sustained Pareto fronts. Zhao *et al.* (2015) discussed a job shop scheduling problem with unrelated parallel machines, and a two-generation Pareto ant colony multi-objective algorithm has been formed that splits the problem into two sub-problems with inheritance relationship and the optimization criteria in two phases.

Problem description containing suitable assumptions, mathematical models, and constraints are developed in Sect. 2. Controlled elitist non-dominated sorting genetic algorithm (CE-NSGA-II) Framework is discussed in Sect. 3. Section 4 consists of a demonstrative example for three cases. The results obtained are discussed in Sect. 5. Section 6 includes the conclusion of the paper and discussion on the scope for future opportunities.

2 Problem Description

A problem of the distributed network-based manufacturing system is considered, which is having a set of n jobs $\{J_1, \ldots, J_n\}$ of orders accepted from various customers, and m available machines $\{R_1, \ldots, R_m\}$. A particular set of schemes or substitute process plans are associated with each job J_i and a series of sequential V_i operations ($O_{il}, \ldots, O_{ivi}$) is linked to each process plan. Consequently, the available machines are employed to process jobs with different possible process plans at different enterprises to achieve better use of resources and satisfactory delivery schedules. Each machine can be operated with different speeds due to their dispersion over geographically distributed enterprises, thus each task is associated with an integer energy E_{Oil} and duration P_{Oil} used by the corresponding machine. The association between duration and energy can be expressed as "Job (Speed) || Makespan, Energy" for this problem. Each task is operated with altered speeds and each speed results in a specific processing time and energy consumption. The operation time decreases with the increase in working speed which also results in increased energy consumption. In this paper, different jobs, their predecessor and successor operations, machine candidates, processing time, energy efficiency have been considered. The objective is to determine the best suitable enterprise and feasible schedule which combines minimization of the makespan and the energy consumed by the machines with maximized resource utilization. Hence, the dead time can be retrieved by escalating the machine's speed if a task is delayed to recover the original solution. The mentioned network-based manufacturing case is one of the intricate problems of the current scenario where a vital role is played by the servicing operating time and transporting time between two corresponding machines to sustain process planning and scheduling functions. Being a large research space, the problem becomes intricate for resulting out the best suiting optimal solution. Therefore, IPPS provides the prospective to produce an effectual optimized process plan due to flexibility in networked manufacturing. Hence, the origination of optimal process plans related to every job linked with constraints becomes a demanding task so; it can be taken as a new problem on the report of current manufacturing circumstances.

2.1 Assumptions

1. Job pre-emption is not permitted.
2. Any task of a job being operated on any available machine should be completed without any interruption until it finishes.
3. Only one job is to be handled by each machine at a point time.

4. All jobs and machines can go under process on-time zero.
5. A linear relation between energy consumption and speed is considered.
6. A linear relation between energy consumption and speed is considered.

2.2 Decision Variables

$$\alpha_{jp} = \begin{cases} 1; & \text{if the } p^{th} \text{ flexible process plan of } j^{th} \text{ job is selected} \\ 0; & \text{otherwise} \end{cases}$$

$$\beta_{jpoQrsm} = \begin{cases} 1; & \text{the operation } U_{jpo} \text{ precedes the operation } O_{Qrs} \text{ on machine m} \\ 0; & \text{otherwise} \end{cases}$$

$$\gamma_{jpom} = \begin{cases} 1; & \text{if machine } m \text{ is selected for } O_{jpo} \\ 0; & \text{otherwise} \end{cases}$$

$$A_{jpom} = \begin{cases} 1, & \text{if the operation } O_{jpo} \text{ is being processed on } m^{th} \text{ machine} \\ 0, & \text{otherwise} \end{cases}$$

$$G_{jpoqrkm} = \begin{cases} 1 & \text{if operation } O_{jpo} \text{ is the successor of operation } O_{qrk} \text{ on } m^{th} \text{ machine} \\ 0 & \text{if operation } O_{jpo} \text{ and operation } O_{qrk} \text{ are not adjacent} \\ -1 & \text{if operation } O_{jpo} \text{ is preceding operation } O_{qrk} \end{cases}$$

$$D_{jpom} = \begin{cases} 1, & \text{if } m^{th} \text{ machine is to be turned off between the operation } O_{jpo} \text{ and operation } O_{qrk} \\ 0, & \text{if } m^{th} \text{ machine to be turned on between the operation } O_{jpo} \text{ and operation } O_{qrk} \end{cases}$$

2.3 Mathematical Modelling

The desired outcomes of this problem are minimization of makespan, maximization of machine utilization, and minimization of energy consumption as represented as follows:

Objectives:

Makespan minimisation:

$$T_j = \text{Max } C_{jpom} \tag{1}$$

Maximization of machine utilization:

$$mu = \frac{\sum_{s=1}^{m} mpt_s}{\sum_{s=1}^{m} (mct_s - mst_s)} \tag{2}$$

Minimization of energy consumption:

$$\emptyset = \sum_{m=1}^{S} (\emptyset_{0/1m} + \emptyset_{im} + \emptyset_{pm} + \emptyset_{sm}) \tag{3}$$

2.3.1 The Energy Consumption Model

It can be divided into four operating states as follows.

a. The turning on/off state

$$\emptyset_{0/1m} = z_m \left[\int_{F_m}^{F_m + D_{m1}} \mu_m(t)dt + \int_{G_m - D_{m2}}^{G_m} \mu_m(t)dt \right] \max_{j,o}(X_{ijkq}) \tag{4}$$

$$F_m = \min_{j,o}(A_{jpom}(E_{jpom} - Pt_{jpom})) \tag{5}$$

$$G_m = \max_{j,o}(A_{jpom} E_{jpom}) \tag{6}$$

b. The idle state

$$\emptyset_{im} = \sum_{j,q}^{N} \sum_{o,k}^{\max(V_{jp}, V_{qr})} \emptyset_{imjpoqrk} \tag{7}$$

$$\emptyset_{imjpoqrk} = \begin{cases} Q(Q_1(\max(F_{m_1} + L_m, E_{jpom}) - E_{jpom}) + (Q_2(\max(F_{m_1} + L_m, E_{qrkm}) - E_{qrkm}), D_{jpoqrk} = 1 \\ Q(Q_1(E_{qrkm} - Pt_{qrkm} - E_{jpom}) + Q_2(E_{jpom} - Pt_{jpom} - E_{qrkm})), D_{jpoqrk} = 0 \end{cases} \tag{8}$$

Where $Q = A_{jpom}A_{qrkm}(G_{jpoqrkm}/2)$, $Q_1 = I_m(G_{jpoqrkm} - 1)$ and $Q_2 = I_m(G_{jpoqrkm} + 1)$

$$F_{m1} = \begin{cases} \min_{a,b}(E_{aebm} - Pt_{aebm})A_{aebm}A_{jpom}A_{qrkm}, D_{aebqrk} + U_{aebjpo} = 1 \cap U_{jpoqrk} = 1 \cap 0 < E_{aebm} < K \\ F_m, \textit{ otherwise} \end{cases} \tag{9}$$

$$K = (G_{jpoqrkm}/2)(G_{jpoqrkm} + 1)(E_{qrkm} - Pt_{qrkm}) + (G_{jpoqrkm} - 1)(E_{jpom} - Pt_{jpom}) \tag{10}$$

c. The processing state

$$\emptyset_{pm} = W_m \left(\sum_{j}^{N} \sum_{o}^{V_{jp}} A_{jpom} Pt_{jpom} \right) \tag{11}$$

d. The standby state

$$\emptyset_{sm} = C_{\max} Y_{sm} \tag{12}$$

Constraints:

For the initial operation in the pth process plan for j^{th} job:

$$C_{jp1m} + \theta(1 - \alpha_{jp}) \geq Pt_{jp1m} \tag{13}$$

For the final operation in the pth process plan for j^{th} job:

$$C_{jv_{jp}jp} - \theta(1 - \alpha_{jp}) \leq C_{jpom} \tag{14}$$

A job's different operations cannot be processed simultaneously

$$C_{jpom} - C_{jp(o-1)m} + \theta(1 - \alpha_{jp}) \geq Pt_{jpom} \tag{15}$$

Only one job is to be handled by each machine at a point of time

$$C_{jpom} - C_{Qrsm} + \theta\beta_{jpoQrsm} \geq Pt_{jpom} \tag{16}$$

Each job can be associated with single process plan

$$\sum_{p=1}^{P_j} \alpha_{jp} = 1 \tag{17}$$

Each operation can be process on a single machine only

$$\sum_{m=1}^{S} \gamma_{jpom} = 1 \tag{18}$$

$$E_{jpo} - Pt_{jpo} \geq E_{jp(o-1)} \tag{19}$$

$$\begin{aligned}&(G_{jpoqrkm}/2)(G_{jpoqrkm} - 1)(E_{qrk} - E_{jpo} - Pt_{qrkm})A_{jpom}A_{qrkm}\\&+ (G_{jpoqrkm}/2)(G_{jpoqrkm} + 1)(E_{jpo} - E_{qrk} - Pt_{jpom})A_{jpom}A_{qrkm} \geq 0\end{aligned} \tag{20}$$

$$\sum_{m=1}^{S} A_{jpom} = 1 \tag{21}$$

$$E_{jpom} \geq 0 \tag{22}$$

$$Pt_{jpom} \geq 0 \tag{23}$$

This problem's objectives are to primarily emphasize on job scheduling so that maximum of their total completion time of all operations can be minimized, i.e., makespan as specified in Eq. (1); maximization of the machine or resource utilization as referred in Eq. (2); and to minimize the Energy consumption of machines as specified in energy consumption framework. The problem is subjected to certain constraints which are listed in the Eqs. (13)–(25). Constraint (13) expresses restriction in processing various operations related to a job concurrently for alternative process plans. Constraint (14) states that only a single operation can be operated on any machine at a time. Constraint (14) comprises that each job can be related to only one single process plan. Constraint (16) shows that a single machine is to be selected for each operation. The precedence relation is depicted in Eq. (19). Constraint (20) states that only one job can be processed by any machine at a point of time. Constraint (21) represents a decision variable that is used to assign availability of the machines at a particular time. Practicality of our problem is portrayed in (22) and (23).

3 Controlled Elitist Non Dominated Sorting Genetic Algorithm (CE-NSGA-II)

Quick erasure of non-elitist front solutions and lack of variety in some decision variables are undesired outcomes of NSGA-II, which leads to the significance of the much superior CE-NSGA-II algorithm. The following sections describe various stages of the CE-NSGA-II algorithm in the reference of networked manufacturing problems having multi-objectives and above-disclosed outcomes of NSGA-II and the parameters for the proposed algorithm is shown in Table 1.

Table 1. Parameter values for algorithm

Parameter	Parameter value
Population size	50
Maximum generations	50
Mutation probability (M_p)	0.05
Cross-over probability (C_p)	0.84

3.1 Population Initialization

At first, the initial population is generated arbitrarily for specified population size. The process also plays a major role in obtaining more optimal solutions. It has been proven that the initial populations generated may affect the best value of the objective function and that their effects may last for several generations (Maaranen *et al.* 2007).

3.2 Evolutionary Operations

In the process of evolution, genetic variation is necessary. The operations that are performed are analogous to those happening in nature in the process of evolution: Survival of the fittest or selection, mutation, and crossover (also called as reproduction or recombination). These operations are carried out to protect the diversity in the population *N* and to create a new child population of the same size. The detailed flowchart of the CE-NSGA-II is detailed in Fig. 1.

Selection. In this operation, better solutions are given more preference which allows them to pass on their genes to the next generation during the execution of the algorithm. This is the preliminary step before performing cross-over or mutation. The best solutions are selected using the fitness values of the objective functions. The fitness value represents the closeness of the solution in achieving the specified objective. There are various methods of selection used for different applications. In our case, we have used a tournament selection in the algorithm.

Mutation. The mutation is used to develop the solution space by generating neighbors in different directions. It can be executed in four methods as swap, insertion, displacement, and inversion where the first two produce close neighbors while the other two create distant neighbors. This operation protects the robust intermediate solutions by diversification and to adjust the fragile ones. Here bit wise mutation is employed with Pm (mutation probability) as 0.05.

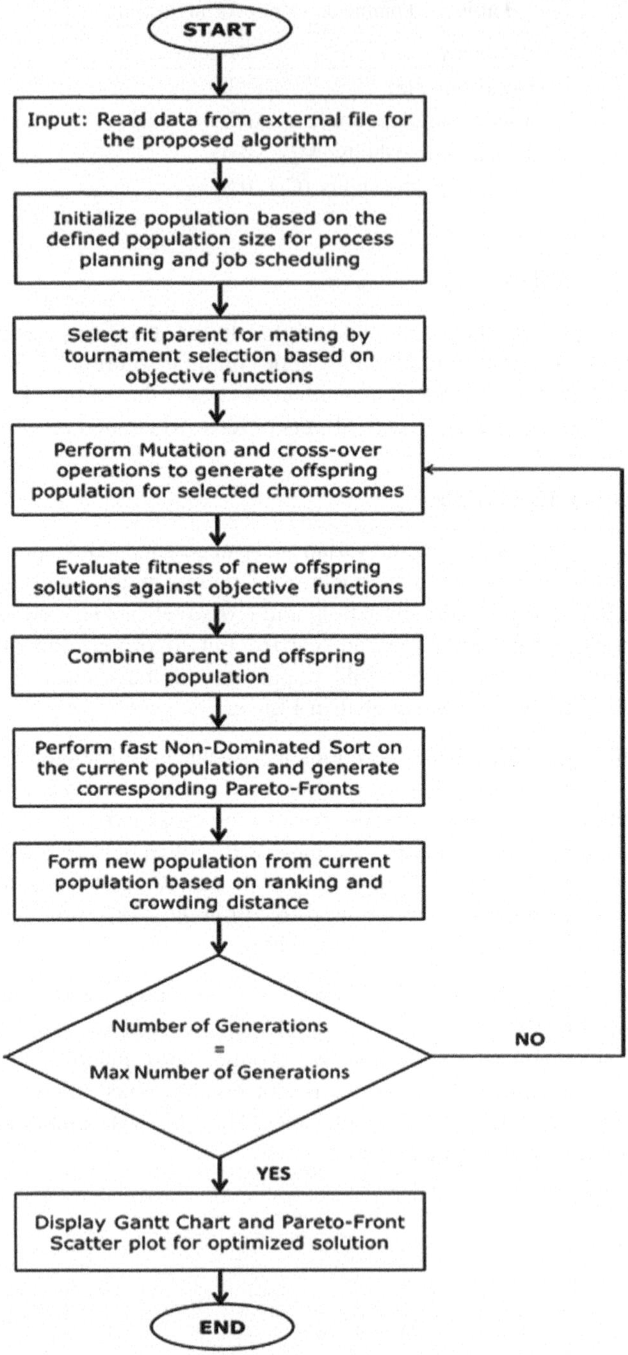

Fig. 1. CE-NSGA-II framework

4 Demonstrative Example

Three illustrative instances (represented by n × m/Job X Machine problem) have been chosen as a testbed to represent the performance and efficacy of the proposed algorithm. Available machine and processing time data are taken (Zhou et al. 2010; Shao et al. 2009) and the energy consumption is computed through the mathematical model using practical machine power and speed parameters. CE-NSGA-II algorithm is applied to obtain optimized process planning and scheduling with minimum makespan, maximum machine utilization, and minimum energy consumption. Tables 1, 2, and 3 have the input data to the algorithm for the three cases respectively. Three different brackets are used to represent the parameter's data. Available machines, processing time, and energy consumption are placed in curly, square, and round brackets respectively. For example, j_3 job's O_3 has available machines as {2, 5}, corresponding [5, 6] minutes processing time, and (372, 378) kilo-joule energy consumption.

4.1 Scenario 1

This scenario exhibits the order placed by costumers for six different jobs where the corresponding operations are to be processed by six geographically located machines at different enterprises. Anyone of multiple process plans can be employed for each job and multiple machines are available to process each operation. Popularly applied scheduling representation means i.e., Gantt chart is characterized in Fig. 2, which pictures the allotment of operations on various available machines, beginning and ending time of each operation, and the optimized process plan. Time and machines are represented on *X* and *Y*-axis of the Gantt chart. The output results for makespan selected optimal process plan and job scheduling obtained from the algorithm are shown in Table 4.

4.2 Scenario 2

Case of eight different jobs from clients to be processed at eight machines available in the network manufacturing system is explored in scenario 2. The output results for makespan selected optimal process plan and job scheduling obtained from the algorithm are shown in Table 5. The job scheduling Gantt chart is presented in Fig. 3.

4.3 Scenario 3

This scenario consists of six different jobs to be completed through processing operations using eight available machines. The output results for makespan selected optimal process plan and job scheduling obtained from the algorithm are shown in Table 6 and the Gantt chart is shown in Fig. 4.

Table 2. Input data of scenario 1

Input data of scenario 1							
		Operations					
Job	PP	O_1	O_2	O_3	O_4	O_5	O_6
J_1	$PP_{1,1}$	{1, 2}	{3, 4, 5}	{6}			
		[6, 5]	[7, 6, 6]	[8]			
		(493, 372)	(496, 410, 378)	(619)			
	$PP_{1,2}$	{1, 3}	{2, 4}	{3, 5}	{4, 5, 6}		
		[4, 5]	[4, 5]	[5, 6]	[5, 5, 4]		
		(329, 354)	(298, 342)	(354, 378)	(342, 315, 310)		
J_2	$PP_{2,1}$	{2}	{1, 3}	{2, 4, 6}	{3, 5}	{2, 4}	{4, 6}
		[4]	[2, 3]	[4, 3, 5]	[2, 4]	[3, 4]	[3, 5]
		(298)	(164, 212)	(298, 205, 387)	(142, 252)	(223, 274)	(205, 387)
	$PP_{2,2}$	{1, 3, 5}	{4}	{4, 6}	{4}	{4, 6}	{1, 6}
		[1, 5, 7]	[5]	[1, 6]	[4]	[1, 2]	[5, 6]
		(82, 354, 441)	(342)	(68, 464)	(274)	(68, 155)	(411, 464)
J_3	$PP_{3,1}$	{2, 3}	{1, 4}	{2, 5}	{3, 6}	{1, 6}	{5}
		[5, 6]	[6, 5]	[5, 6]	[6, 5]	[6, 6]	[4]
		(372, 425)	(493, 342)	(372, 378)	(425, 387)	(493, 464)	(252)
	$PP_{3,2}$	{1}	{3, 4}	{5}			
		[7]	[8, 8]	[9]			
		(575)	(566, 547)	(567)			
	$PP_{3,3}$	{2, 3}	{4}	{3, 5}	{4, 6}	{1, 2}	
		[7, 6]	[7]	[7, 8]	[7, 8]	[1, 4]	
		(521, 425)	(479)	(496, 504)	(479, 619)	(82, 298)	
J_4	$PP_{4,1}$	{1, 2}	{3, 4}	{6}	{1}		
		[7, 8]	[7, 6]	[9]	[5]		
		(575, 595)	(496, 410)	(697)	(411)		
	$PP_{4,2}$	{1, 3, 5}	{2}	{3, 4, 6}	{5, 6}		
		[4, 3, 7]	[4]	[4, 5, 6]	[3, 5]		
		(329, 212, 441)	(298)	(283, 342, 464)	(189, 387)		
J_5	$PP_{5,1}$	{1}	{2, 4}	{3}	{5, 6}		
		[3]	[4, 4]	[4]	[3, 3]		
		(247)	(298, 274)	(283)	(189, 232)		
	$PP_{5,2}$	{2, 4}	{5}	{3, 6}			
		[5, 6]	[7]	[9, 8]			
		(372, 410)	(441)	(437, 619)			
J_6	$PP_{6,1}$	{1, 2}	{3, 4}	{2, 5}	{3}	{4, 5}	{3, 6}
		[3, 4]	[4, 3]	[5, 3]	[4]	[5, 6]	[5, 4]
		(247, 298)	(283, 205)	(372, 189)	(283)	(343, 378)	(354, 310)
	$PP_{6,2}$	{1, 3}	{2, 3}	{2, 4}	{6}		
		[4, 4]	[5, 6]	[6, 7]	[7]		
		(329, 283)	(372, 425)	(446, 479)	(542)		
	$PP_{6,3}$	{1, 2, 3}	{4, 5}	{3, 6}			
		[3, 5, 8]	[7, 10]	[9, 9]			
		(247, 372, 566)	(479, 630)	(637, 697)			

Table 3. Input data of scenario 2

Input data of scenario 2						
		Operations				
Job	PP	O_1	O_2	O_3	O_4	O_5
J_1	$PP_{1,1}$	{2, 4}	{7, 8}	{1, 2}	{8, 6}	{3, 8}
		[18, 22]	[39, 36]	[11, 10]	[31, 34]	[26, 24]
		(1339, 1505)	(3346, 2138)	(904, 744)	(1841, 2632)	(1841, 1426)
	$PP_{1,2}$	{2, 4}	{3, 5}	{1, 2, 4}	{5, 6}	{3, 8}
		[18, 22]	[21, 23]	[10, 12, 15]	[32, 30]	[26, 24]
		(1339, 1505)	(1487, 1449)	(822, 893, 1026)	(2016, 2322)	(1841, 1426)
	$PP_{1,3}$	{2, 4}	{3, 5}	{1, 7}	{3, 8}	
		[18, 22]	[21, 23]	[45, 44]	[26, 24]	
		(1339, 1505)	(1487, 1449)	(3699, 3775)	(1841, 1426)	
J_2	$PP_{2,1}$	{2, 4}	{7, 8}	{1, 2}	{3, 8}	
		[18, 22]	[39, 36]	[37, 39]	[26, 24]	
		(1339, 1505)	(3346, 2138)	(3041, 2902)	(1842, 1426)	
	$PP_{2,2}$	{2, 4}	{8, 6}	{1, 2, 4}	{5, 6}	{3, 8}
		[18, 22]	[20, 21]	[10, 12, 15]	[36, 38]	[26, 24]
		(1339, 1505)	(1188, 1625)	(822, 893, 1026)	(2268, 2941)	(1841, 1426)
	$PP_{2,3}$	{2, 4}	{8, 6}	{1, 7}	{3, 8}	
		[18, 22]	[20, 21]	[45, 44]	[26, 24]	
		(1339, 1505)	(1188, 1625)	(3699, 3775)	(1842, 1426)	
	$PP_{2,4}$	{2, 4}	{3, 5}	{1, 2, 4}	{5, 6}	{3, 8}
		[18, 22]	[21, 23]	[10, 12, 15]	[36, 38]	[26, 24]
		(1339, 1505)	(1487, 1449)	(822, 893, 1026)	(2268, 2941)	(1841, 1426)
	$PP_{2,5}$	{2, 4}	{3, 5}	{1, 7}	{3, 8}	
		[18, 22]	[21, 23]	[45, 44]	[26, 24]	
		(1339, 1505)	(1487, 1449)	(3699, 3775)	(1842, 1426)	
J_3	$PP_{3,1}$	{1, 4}	{6, 7}	{5, 8}	{2, 4}	
		[22, 25]	[24, 22]	[20, 19]	[22, 27]	
		(1808, 1710)	(1858, 1888)	(1260, 1129)	(1637, 1847)	
	$PP_{3,2}$	{3, 5}	{4, 6}	{2, 3, 5}	{2, 4}	
		[12, 15]	[24, 23]	[30, 31, 24]	[22, 27]	
		(850, 945)	(1642, 1780)	(2232, 2195, 1512)	(1637, 1847)	
	$PP_{3,3}$	{3, 5}	{6, 7}	{1, 8}	{2, 4}	
		[12, 15]	[21, 22]	[32, 30]	[22, 27]	
		(850, 945)	(1625, 1888)	(2630, 1782)	(1637, 1847)	
J_4	$PP_{4,1}$	{1, 4}	{6, 7}	{2, 4}		
		[22, 25]	[42, 44]	[22, 27]		
		(1808, 1710)	(3251, 3775)	(1637, 1847)		
	$PP_{4,2}$	{1, 4}	{5, 8}	{2, 4}		
		[22, 25]	[41, 43]	[22, 27]		

(*continued*)

Table 3. (*continued*)

Input data of scenario 2						
		Operations				
Job	PP	O_1	O_2	O_3	O_4	O_5
		(1808, 1710)	(2583, 2554)	(1637, 1847)		
	$PP_{4,3}$	{3, 5}	{4, 5}	{2, 3, 5}	{2, 4}	
		[12, 15]	[24, 23]	[30, 31, 29]	[22, 27]	
		(850, 945)	(1642, 1449)	(2232, 2195, 1827)	(1637, 1847)	
	$PP_{4,4}$	{3, 5}	{6, 7}	{1, 8}	{2, 4}	
		[12, 15]	[21, 22]	[32, 30]	[22, 27]	
		(850, 945)	(1625, 1888)	(2630, 1782)	(1637, 1847)	
J_5	$PP_{5,1}$	{2, 4}	{1, 3}	{6, 7}	{5, 8}	{3, 4}
		[18, 22]	[22, 25]	[24, 22]	[20, 18]	[22, 27]
		(1339, 1505)	(1808, 1770)	(1858, 1888)	(1260, 1069)	(1558, 1847)
	$PP_{5,2}$	{2, 4}	{3, 5}	{6, 8}	{1, 7}	{3, 4}
		[18, 22]	[12, 15]	[19, 21]	[32, 31]	[22, 27]
		(1339, 1505)	(850, 945)	(1471, 1247)	(2630, 2660)	(1558, 1847)
	$PP_{5,3}$	{2, 4}	{3, 5}	{1, 2, 6}	{3, 4}	
		[18, 22]	[12, 15]	[50, 52, 54]	[22, 27]	
		(1339, 1505)	(850, 945)	(4110, 3869, 4180)	(1558, 1847)	
J_6	$PP_{6,1}$	{2, 4}	{1, 7}	{6, 7}	{5, 8}	{3, 4}
		[18, 22]	[22, 24]	[24, 22]	[20, 18]	[22, 27]
		(1339, 1505)	(1808, 2059)	(1858, 1888)	(1260, 1069)	(1558, 1847)
	$PP_{6,2}$	{2, 4}	{1, 3}	{6, 7}	{5, 8}	{3, 4}
		[18, 22]	[21, 25]	[24, 22]	[20, 18]	[22, 27]
		(1339, 1505)	(1726, 1770)	(1858, 1888)	(1260, 1069)	(1558, 1847)
	$PP_{6,3}$	{2, 4}	{3, 5}	{6, 8}	{3, 4}	
		[18, 22]	[12, 15]	[53, 51]	[22, 27]	
		(1339, 1505)	(850, 945)	(4102, 3029)	(1558, 1847)	
	$PP_{6,4}$	{2, 4}	{3, 5}	{1, 2, 6}	{3, 4}	
		[18, 22]	[12, 15]	[50, 52, 54]	[22, 27]	
		(1339, 1505)	(850, 945)	(4110, 3869, 4180)	(1558, 1847)	

Table 4. Input data of scenario 3

Input data of scenario 3						
		Operations				
Job	PP	O_1	O_2	O_3	O_4	O_5
J_1	$PP_{1,1}$	{2, 4}	{7, 8}	{1, 2}	{8, 6}	{3, 8}
		[18, 22]	[39, 36]	[11, 10]	[31, 34]	[26, 24]
		(1339, 1505)	(3346, 2138)	(904, 744)	(1841, 2632)	(1841, 1426)
	$PP_{1,2}$	{2, 4}	{3, 5}	{1, 2, 4}	{5, 6}	{3, 8}

(*continued*)

Table 4. (*continued*)

Input data of scenario 3						
		Operations				
Job	PP	O_1	O_2	O_3	O_4	O_5
		[18, 22]	[21, 23]	[10, 12, 15]	[32, 30]	[26, 24]
		(1339, 1505)	(1487, 1449)	(822, 893, 1026)	(2016, 2322)	(1841, 1426)
	$PP_{1,3}$	{2, 4}	{3, 5}	{1, 7}	{3, 8}	
		[18, 22]	[21, 23]	[45, 44]	[26, 24]	
		(1339, 1505)	(1487, 1449)	(3699, 3775)	(1841, 1426)	
J_2	$PP_{2,1}$	{2, 4}	{7, 8}	{1, 2}	{3, 8}	
		[18, 22]	[39, 36]	[37, 39]	[26, 24]	
		(1339, 1505)	(3346, 2138)	(3041, 2902)	(1842, 1426)	
	$PP_{2,2}$	{2, 4}	{8, 6}	{1, 2, 4}	{5, 6}	{3, 8}
		[18, 22]	[20, 21]	[10, 12, 15]	[36, 38]	[26, 24]
		(1339, 1505)	(1188, 1625)	(822, 893, 1026)	(2268, 2941)	(1841, 1426)
	$PP_{2,3}$	{2, 4}	{8, 6}	{1, 7}	{3, 8}	
		[18, 22]	[20, 21]	[45, 44]	[26, 24]	
		(1339, 1505)	(1188, 1625)	(3699, 3775)	(1842, 1426)	
	$PP_{2,4}$	{2, 4}	{3, 5}	{1, 2, 4}	{5, 6}	{3, 8}
		[18, 22]	[21, 23]	[10, 12, 15]	[36, 38]	[26, 24]
		(1339, 1505)	(1487, 1449)	(822, 893, 1026)	(2268, 2941)	(1841, 1426)
	$PP_{2,5}$	{2, 4}	{3, 5}	{1, 7}	{3, 8}	
		[18, 22]	[21, 23]	[45, 44]	[26, 24]	
		(1339, 1505)	(1487, 1449)	(3699, 3775)	(1842, 1426)	
J_3	$PP_{3,1}$	{1, 4}	{6, 7}	{5, 8}	{2, 4}	
		[22, 25]	[24, 22]	[20, 19]	[22, 27]	
		(1808, 1710)	(1858, 1888)	(1260, 1129)	(1637, 1847)	
	$PP_{3,2}$	{3, 5}	{4, 6}	{2, 3, 5}	{2, 4}	
		[12, 15]	[24, 23]	[30, 31, 24]	[22, 27]	
		(850, 945)	(1642, 1780)	(2232, 2195, 1512)	(1637, 1847)	
	$PP_{3,3}$	{3, 5}	{6, 7}	{1, 8}	{2, 4}	
		[12, 15]	[21, 22]	[32, 30]	[22, 27]	
		(850, 945)	(1625, 1888)	(2630, 1782)	(1637, 1847)	
J_4	$PP_{4,1}$	{1, 4}	{6, 7}	{2, 4}		
		[22, 25]	[42, 44]	[22, 27]		
		(1808, 1710)	(3251, 3775)	(1637, 1847)		
	$PP_{4,2}$	{1, 4}	{5, 8}	{2, 4}		
		[22, 25]	[41, 43]	[22, 27]		
		(1808, 1710)	(2583, 2554)	(1637, 1847)		
	$PP_{4,3}$	{3, 5}	{4, 5}	{2, 3, 5}	{2, 4}	
		[12, 15]	[24, 23]	[30, 31, 29]	[22, 27]	
		(850, 945)	(1642, 1449)	(2232, 2195, 1827)	(1637, 1847)	

(*continued*)

Table 4. (*continued*)

Input data of scenario 3						
		Operations				
Job	PP	O_1	O_2	O_3	O_4	O_5
		{3, 5}	{6, 7}	{1, 8}	{2, 4}	
		[12, 15]	[21, 22]	[32, 30]	[22, 27]	
		(850, 945)	(1625, 1888)	(2630, 1782)	(1637, 1847)	
J_5	$PP_{5,1}$	{2, 4}	{1, 3}	{6, 7}	{5, 8}	{3, 4}
		[18, 22]	[22, 25]	[24, 22]	[20, 18]	[22, 27]
		(1339, 1505)	(1808, 1770)	(1858, 1888)	(1260, 1069)	(1558, 1847)
	$PP_{5,2}$	{2, 4}	{3, 5}	{6, 8}	{1, 7}	{3, 4}
		[18, 22]	[12, 15]	[19, 21]	[32, 31]	[22, 27]
		(1339, 1505)	(850, 945)	(1471, 1247)	(2630, 2660)	(1558, 1847)
	$PP_{5,3}$	{2, 4}	{3, 5}	{1, 2, 6}	{3, 4}	
		[18, 22]	[12, 15]	[50, 52, 54]	[22, 27]	
		(1339, 1505)	(850, 945)	(4110, 3869, 4180)	(1558, 1847)	
J_6	$PP_{6,1}$	{2, 4}	{1, 7}	{6, 7}	{5, 8}	{3, 4}
		[18, 22]	[22, 24]	[24, 22]	[20, 18]	[22, 27]
		(1339, 1505)	(1808, 2059)	(1858, 1888)	(1260, 1069)	(1558, 1847)
	$PP_{6,2}$	{2, 4}	{1, 3}	{6, 7}	{5, 8}	{3, 4}
		[18, 22]	[21, 25]	[24, 22]	[20, 18]	[22, 27]
		(1339, 1505)	(1726, 1770)	(1858, 1888)	(1260, 1069)	(1558, 1847)
	$PP_{6,3}$	{2, 4}	{3, 5}	{6, 8}	{3, 4}	
		[18, 22]	[12, 15]	[53, 51]	[22, 27]	
		(1339, 1505)	(850, 945)	(4102, 3029)	(1558, 1847)	
	$P_{6,4}$	{2, 4}	{3, 5}	{1, 2, 6}	{3, 4}	
		[18, 22]	[12, 15]	[50, 52, 54]	[22, 27]	
		(1339, 1505)	(850, 945)	(4110, 3869, 4180)	(1558, 1847)	

Table 5. Job scheduling of case 1

Job no	Case 1		
	Makespan (Min)	Process plan	Scheduling of jobs
1	20	2	[3, 2, 5, 5]
2	18	2	[1, 4, 4, 4, 4, 6]
3	24	2	[1, 4, 5]
4	22	2	[5, 2, 6, 6]
5	20	2	[2, 5, 6]
6	24	2	[3, 3, 4, 6]

Table 6. Job scheduling of case 2

Job no	Case 2		
	Makespan (Min)	Process plan	Scheduling of jobs
1	22	2	[1, 2, 3, 8]
2	24	2	[3, 4, 4, 4, 4, 8, 4]
3	23	2	[1, 3, 8]
4	19	2	[1, 2, 6, 6]
5	14	1	[1, 2, 3, 8]
6	22	2	[3, 3, 7, 6]
7	23	1	[2, 4, 6, 5, 6]
8	18	2	[2, 4, 6]

Table 7. Job scheduling of case 3

Job no	Case 3		
	Makespan (Min)	Process plan	Scheduling of jobs
1	107	3	[2, 3, 7, 8]
2	109	5	[2, 3, 7, 3]
3	105	2	[5, 6, 2, 4]
4	94	3	[3, 4, 3, 4]
5	116	3	[4, 3, 6, 4]
6	105	4	[2, 5, 1, 3]

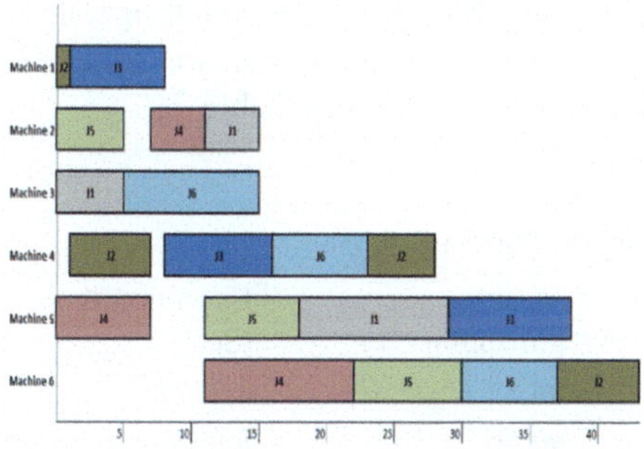

Fig. 2. Gantt chart showing job scheduling for scenario 1

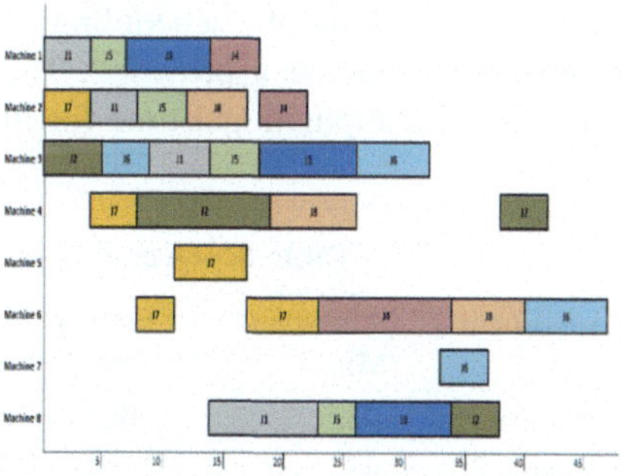

Fig. 3. Gantt chart showing job scheduling for scenario 2

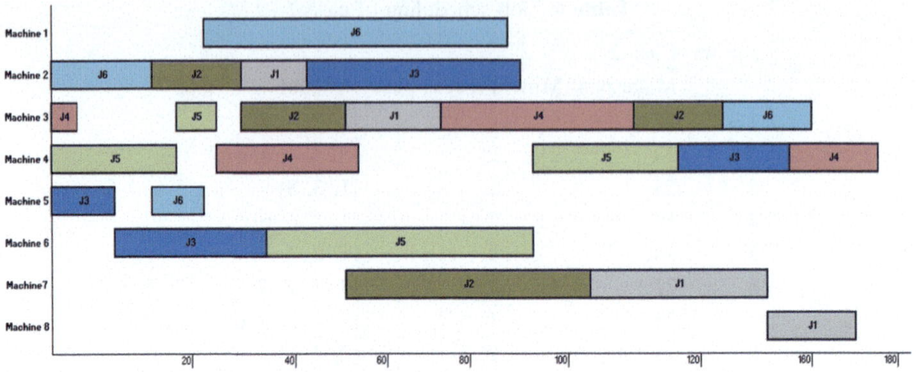

Fig. 4. Gantt chart showing job scheduling for scenario 3

5 Results and Discussion

The parameters deciding the performance of the projected algorithm are the utilization of machines, makespan, number of generations, and computational run time. Three incompatible objectives are selected in our paper as minimization of makespan, minimization of energy consumption, and maximization of machine utilization. Better efficiency of the algorithm and higher productivity of the manufacturing system can be assured with lesser makespan value.

In Tables 4, 5 and 6, the total makespan i.e., the total of processing times on each machine for each job; the optimal process plan that is selected from the set of alternatives provided and the scheduling of jobs as to "on which machine which operation is performed" has been provided. The makespan, energy consumption, machine utilization, and computational time for fittest solutions are detailed in Table 7.

Table 8. Overall optimized value of objective functions

Experiment	Makespan (Min)	Energy consumption (KJ)	Machine utilization (%)	Execution time (Sec)
Scenario 1	43	9083	94.82	4.2377
Scenario 2	47	11889	90.16	5.6243
Scenario 3	173	46142	91.12	5.1098

It can be referred that the proposed algorithm converges fast because it encounters termination norms in a lower number of generations. Though, obtaining lower computational time with fewer generations is not guaranteed because each algorithm has particular intricacy implicated for every single run of programming. This occurrence leads to the concern of computational time and the proposed algorithm deals with several generations and computational time in an equivalent manner and provides justified computational time so that reaction of manufacturing resources is improved through quick process plan generation. Energy efficiency and cost-effectiveness of any manufacturing sector can be improved with lower energy consumption through proper process plan and scheduling selection which can be inferred from our results. One more crucial factor is the superior consumption of manufacturing resources which is termed as "Machine Utilization" and can be defined as the percentage ratio of real operation processing time to the overall running time of all the machines (Table 8).

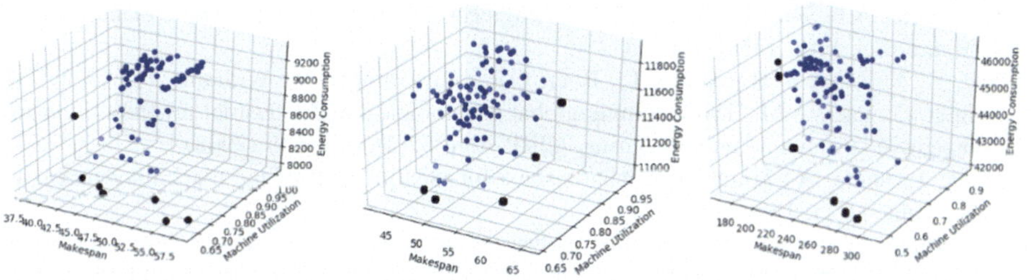

Fig. 5. Pareto fronts for scenario 1, 2 and 3

The scatter plots of the optimal Pareto-fronts for the three cases have been shown in Fig. 5. The dark highlighted points show the locus of the optimal Pareto-fronts obtained. A Pareto-front contains all reasonable solutions based on the objective functions and constraints. Although there are various methods to choose the best solution; one of the simplest and easiest ways is to compare with the ideal point and recognize the point closest to the ideal point. Also, the solution chosen from the frontiers depends on the perspective of the decision-maker. This CE-NSGA-II algorithm has been coded in Python 3 and tested on Intel® Core™ i5-4200U CPU @1.6 GHz 2.30 GHz, 4 GB of RAM. In the end, it can be realized from the results that CE-NSGA-II executes efficiently with a lower number of generations to obtain optimized objective functions with improved convergence and divergence of solutions.

6 Conclusion

In this research study, investigation of variation between traditional and network manufacturing systems helped us to identify the importance and necessity of networked manufacturing system and their characteristics. Consequently, problem description with the mathematical model is formulated with certain assumptions, constraints, and makespan minimization, machine utilization maximization, and minimization of energy consumption are taken as objectives with the IPPS approach. Practical machine parameters are characterized like speed, power, etc. and the energy consumption is computed through a mathematical model for each operation-machine combination. The problem being a type of NP-hard complex problem, the CE-NSGA-II algorithm is proposed to fulfill the objective needs and produce a feasible process plan. Further on need, more sub-objectives are introduced like several generations, each job's optimal process plan and computational time which categorize the performance of the proposed algorithm substantially.

References

Haapala, K.R., Zhao, F., Camelio, J., Sutherland, J.W., Skerlos, S.J., Dornfeld, D.A., Jawahir, I. S., Clarens, A.F., Rickli, J.L.: A review of engineering research in sustainable manufacturing. J. Manuf. Sci. Eng. **135**(4) (2013)

Jovane, F., Yoshikawa, H., Alting, L., Boer, C.R., Westkamper, E., Williams, D., Tseng, M., Seliger, G., Paci, A.M.: The incoming global technological and industrial revolution towards competitive sustainable manufacturing. CIRP Ann. **57**(2), 641–659 (2008)

Peklenik, J., Jerele, A.: Some basic relationships for identification of the machining processes. CIRP Ann. **41**(1), 155–159 (1992)

Veeramani, D., Wang, K.J.: Performance analysis of auction-based distributed shop-floor control schemes from the perspective of the communication system. Int. J. Flex. Manuf. Syst. **9**(2), 121–143 (1997)

De Wilde, P., Briscoe, G.: Stability of evolving multiagent systems. IEEE Trans. Syst. Man Cybern. Part B (Cybern.) **41**(4), 1149–1157 (2011)

Varela, M.L., Putnik, G.D., Manupati, V.K., Rajyalakshmi, G., Trojanowska, J., Machado, J.: Collaborative manufacturing based on cloud, and on other I4. 0 oriented principles and technologies: a systematic literature review and reflections. Manag. Prod. Eng. Rev. **9** (2018)

Ausaf, M.F., Gao, L., Li, X., Al Aqel, G.: A priority-based heuristic algorithm (PBHA) for optimizing integrated process planning and scheduling problem. Cogent Eng. **2**(1), 1070494 (2015)

Lin, J., Liu, M., Hao, J., Jiang, S.: A multi-objective optimization approach for integrated production planning under interval uncertainties in the steel industry. Comput. Oper. Res. **72**, 189–203 (2016)

Luo, G., Wen, X., Li, H., Ming, W., Xie, G.: An effective multi-objective genetic algorithm based on immune principle and external archive for multi-objective integrated process planning and scheduling. Int. J. Adv. Manuf. Technol. **91**(9–12), 3145–3158 (2017)

Zhao, B., Gao, J., Chen, K., Guo, K.: Two-generation Pareto ant colony algorithm for multi-objective job shop scheduling problem with alternative process plans and unrelated parallel machines. J. Intell. Manuf. **29**(1), 93–108 (2015)

Matrix Method for Calculating the Reactions of the Elastic Supports of a Continuous Beam Subjected to the Action of a Load Train

Cornel Marin(✉)

Valahia University of Targoviste, Targoviste, Romania
marin.cornel8@gmail.com

Abstract. Continuous beams are statically indeterminate systems (hyperstatic systems). New analytical methods based on matrix calculation are currently being used to solve such problems, which allow obtaining reactions. Using the MATHCAD professional program you can solve matrix equations and also you can get diagrams of variation of sectional efforts, displacements and rotations of sections along the entire length of the bar. This paper presents the matrix method for solving a particular problem of a beam located on seven prestressed elastic supports for compression.

Keywords: Matrix methods · Continuous beams · Elastic supports

1 Introduction

Continuous beams are statically indeterminate systems (hyperstatic systems) consisting of straight bars supported on more than three rigid or elastic supports, at the same level as the axis of the bar or displaced from the initial straight elastic line of the bar. To solve such problems, matrix methods are currently used that allow obtaining reactions, diagrams of variation of sectional efforts, arrows (displacements along the Oz axis) and rotations of sections (along the Oy axis) using the professional mathematical program MATHCAD.

2 Defining the Problem

It is considered the continuous beam (a straight bar) located on seven elastic supports, which are formed by pairs of springs with stiffness k, prestressed; the continuous beam is subjected to the action of a load train (a force P having a variable position on the bar given by the distance y) and has a constant bending stiffness EI along its entire length; before loading with the load train the elastic supports are all located at the same level with the axis of the bar (Fig. 1). The arrows in the sections corresponding to the elastic supports (displacements along the Oz axis) are proportional to the respective reactions: $w_1 = \frac{V_1}{k}$; $w_2 = \frac{V_2}{k}$; .... $w_7 = \frac{V_7}{k}$.

G. I. Gheorghe (Ed.): ICOMECYME 2020, LNNS 143, pp. 207–214, 2020.
https://doi.org/10.1007/978-3-030-53973-3_22

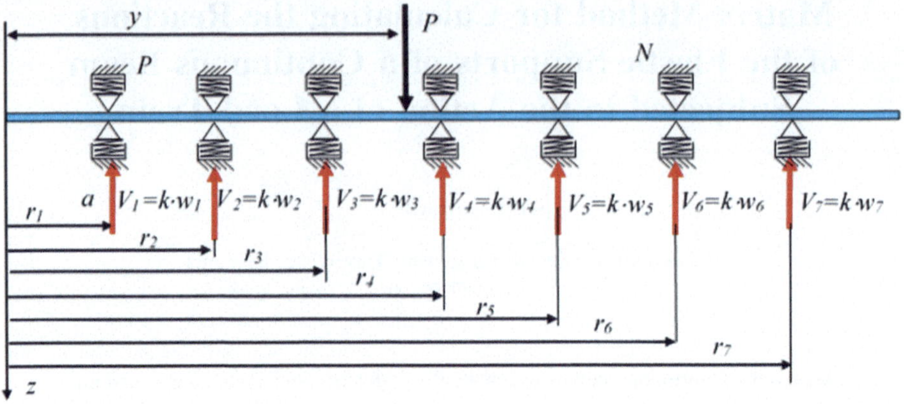

Fig. 1.

During P-train loading, some reactions that are positive (have the opposite direction to the Oz axis) and some become negative (have the direction of the Oz axis). The following question arises: What must be the minimum value of the prestressing of the elastic elements in the supports (by compression) so that they maintain their state of compression throughout the movement of the load train?

3 Matrix Method of Solving

To determine the seven unknown reactions (V1, V2 … V7) write the two equilibrium equations in *Rigid Solid Mechanics*:

$$\begin{aligned} P &= V_1 + V_2 + V_3 + V_4 + V_5 + V_6 + V_7 \\ P \cdot (r_7 - y) &= V_1 \cdot (r_7 - r_1) + V_2 \cdot (r_7 - r_2) + V_3 \cdot (r_7 - r_3) \\ &\quad + V_4 \cdot (r_7 - r_4) + V_5 \cdot (r_7 - r_5) + V_6 \cdot (r_7 - r_6) \end{aligned} \tag{1}$$

The other five equations are obtained from the deformation conditions using the *equation of the three arrows* [1] for the support triplets: *1-2-3; 2-3-4; 3-4-5; 4-5-6 și 5-6-7*:

$$\begin{aligned} EI \cdot [w_1 \cdot d_{23} - w_2 \cdot d_{13} + w_3 \cdot d_{12}] &= \Psi_1 \cdot d_{23} - \Psi_2 \cdot d_{13} + \Psi_3 \cdot d_{12} \\ EI \cdot [w_2 \cdot d_{34} - w_3 \cdot d_{24} + w_4 \cdot d_{23}] &= \Psi_2 \cdot d_{34} - \Psi_3 \cdot d_{24} + \Psi_4 \cdot d_{23} \\ EI \cdot [w_3 \cdot d_{45} - w_4 \cdot d_{35} + w_5 \cdot d_{34}] &= \Psi_3 \cdot d_{45} - \Psi_4 \cdot d_{35} + \Psi_5 \cdot d_{34} \\ EI \cdot [w_4 \cdot d_{56} - w_5 \cdot d_{46} + w_6 \cdot d_{45}] &= \Psi_4 \cdot d_{56} - \Psi_5 \cdot d_{46} + \Psi_6 \cdot d_{45} \\ EI \cdot [w_5 \cdot d_{67} - w_6 \cdot d_{57} + w_7 \cdot d_{56}] &= \Psi_5 \cdot d_{67} - \Psi_6 \cdot d_{57} + \Psi_7 \cdot d_{56} \end{aligned} \tag{2}$$

where: d_{ij} represents the distance between support i and support j:

$$d_{ij} = r_j - r_i \tag{3}$$

Ψ_1, Ψ_2, Ψ_3 .... Ψ_7 are the *loading functions* [1] corresponding to the supports 1, 2, 3, … 7 which depend on the given force P as well as on the unknown reactions V_1, V_2 … V_7:

$$\begin{cases} \Psi_1 = \frac{P\cdot(r_1-y)^3}{6} \\ \Psi_2 = \frac{P\cdot(r_2-y)^3}{6} - \frac{V_1\cdot d_{12}^3}{6}; \\ \Psi_3 = \frac{P\cdot(r_3-y)^3}{6} - \frac{V_1\cdot d_{13}^3}{6} - \frac{V_2\cdot d_{23}^3}{6}; \\ \Psi_4 = \frac{P\cdot(r_4-y)^3}{6} - \frac{V_1\cdot d_{14}^3}{6} - \frac{V_2\cdot d_{24}^3}{6} - \frac{V_3\cdot d_{34}^3}{6} \\ \Psi_5 = \frac{P\cdot(r_5-y)^3}{6} - \frac{V_1\cdot d_{15}^3}{6} - \frac{V_2\cdot d_{25}^3}{6} - \frac{V_3\cdot d_{35}^3}{6} - \frac{V_4\cdot d_{45}^3}{6} \\ \Psi_6 = \frac{P\cdot(r_6-y)^3}{6} - \frac{V_1\cdot d_{16}^3}{6} - \frac{V_2\cdot d_{26}^3}{6} - \frac{V_3\cdot d_{36}^3}{6} - \frac{V_4\cdot d_{46}^3}{6} - \frac{V_5\cdot d_{56}^3}{6} \\ \Psi_7 = \frac{P\cdot(r_7-y)^3}{6} - \frac{V_1\cdot d_{17}^3}{6} - \frac{V_2\cdot d_{27}^3}{6} - \frac{V_3\cdot d_{37}^3}{6} - \frac{V_4\cdot d_{47}^3}{6} - \frac{V_5\cdot d_{57}^3}{6} - \frac{V_6\cdot d_{67}^3}{6} \end{cases} \tag{4}$$

To solve this system of seven equations with seven unknowns, the functions are introduced: *kP1(x, y), k1(x), k2(x),…. k7(x)* using the MATHCAD step function: $\Phi(x - a)$ as in shown in the screenshot in Fig. 2.

Enter the relationships between the arrows in Eqs. (2) w_1, w_2, ... w_7 and reaction V_1, V_2 … V_7:

$$w_1 = \frac{V_1}{k}; \quad w_2 = \frac{V_2}{k}; \ldots; \quad w_7 = \frac{V_7}{k} \tag{5}$$

The ratio between the stiffness of the bar EI and the stiffness k of the springs was denoted by β:

$$\beta = \frac{EI}{k} \tag{6}$$

Substituting in Eq. (1) and (2) we obtain a system of equations that can be written matrix:

$$\boldsymbol{A} \cdot \boldsymbol{X} = \boldsymbol{B}(y) \tag{7}$$

Calculate the inverse matrix of the $\boldsymbol{A}^{-1}$ structure.

Matrix X of unknown reactions $V_1(y)$, $V_2(y)$, $V_3(y)$, *și* $V_4(y)$, is written:

$$\boldsymbol{X}(y) = \boldsymbol{A}^{-1} \cdot \boldsymbol{B}(y) \tag{8}$$

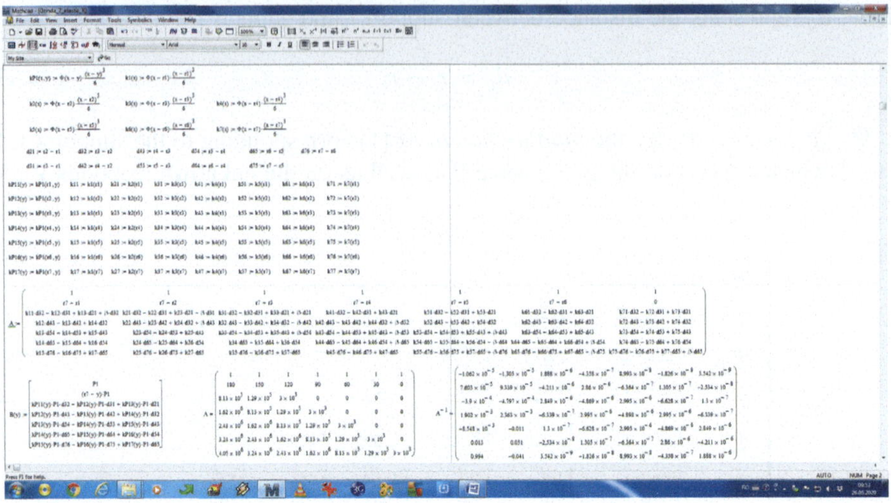

Fig. 2. Screenshot of MATHACD step functions

4 Particular Numerical Case

4.1 Calculation of Reactions to the Movement of the Load Train

In order to determine the variation of the seven unknown reactions as a function of distance y, the particular values of the equidistant elastic beam configuration are considered below:

$$r1 := 10 \quad r2 := 40 \quad r3 := 70 \quad r4 := 100 \quad r5 := 130 \quad r6 := 160 \quad r7 := 190$$

$$P1 := 100\,\mathrm{KN}$$

$$EI := 10000 \quad k := 100\,\mathrm{N/m} \quad \beta := \frac{EI}{k} \quad \beta = 100$$

$$d21 := r2 - r1 \quad d32 := r3 - r2 \quad d43 := r4 - r3 \quad d54 := r5 - r4$$

$$d31 := r3 - r1 \quad d42 := r4 - r2 \quad d53 := r5 - r3 \quad d64 := r6 - r4$$

$$d65 := r6 - r5 \quad d76 := r7 - r6$$

$$d75 := r7 - r5$$

Figures 3 and 4 showed the variations of the reactions $V_1(y)$, $V_2(y)$, $V_3(y)$, *and* $V_4(y)$, the other reactions $V_5(y)$, $V_6(y)$ *and* $V_7(y)$, have a symmetrical variation with the reactions: $V_3(y)$, $V_2(y)$ respectively $V_1(y)$.

The positive values correspond to compressive forces in the elastic elements, while the negative values correspond to tensile forces in the elastic elements.

It is observed that for the value of the relative stiffness factor $\beta = EI/k = 10$ the ratio between positive and negative values is approximately 100 kN/10 kN = 10 if we do not consider the reaction V_1 when the force P is in the console.

For this particular case, the minimum value of the spring prestressing force corresponds to a percentage of about 10% of the maximum value of the force P.

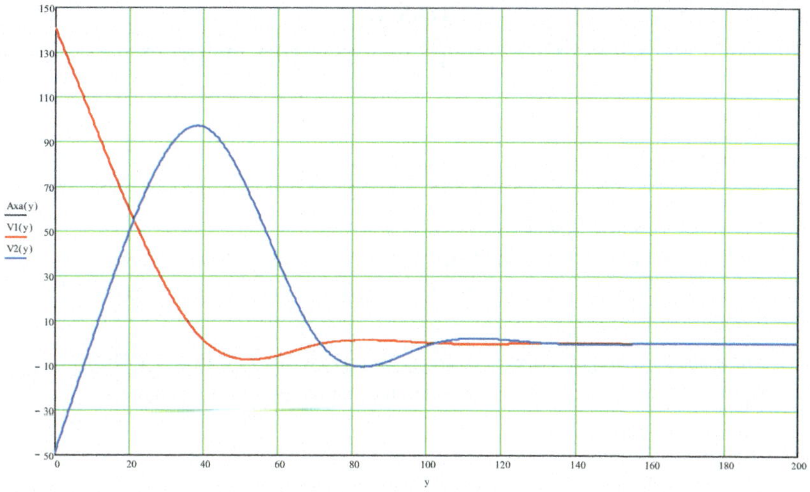

Fig. 3. Diagram of variation of reactions V1 and V2 with distance y (β = 10)

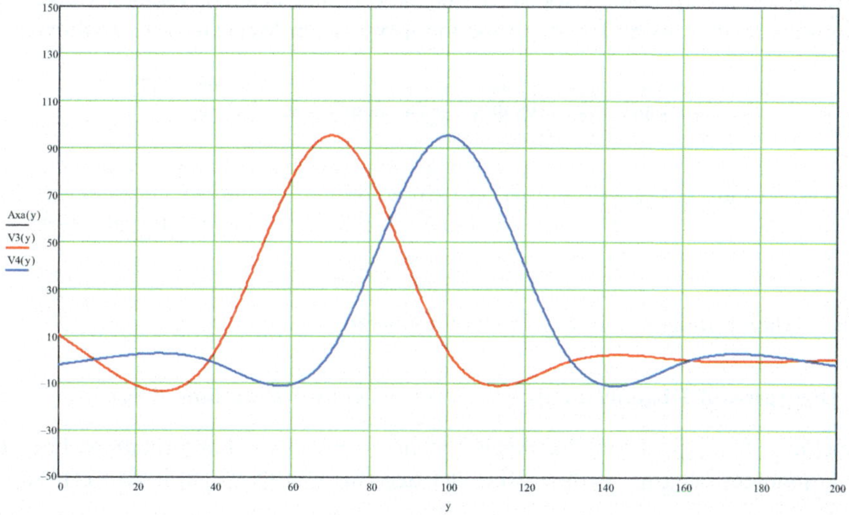

Fig. 4. Diagram of variation of reactions V3 and V4 with distance (β = 10)

In order to study the influence of the factor β on the prestressed force, two values for the relative rigidity are considered $\beta = 1$ și $\beta = 1000$. It is observed that the values in Table 1 are obtained, the same reaction values for all four cases of relative stiffness, differing only the values of the arrows $w_j = V_j/k$.

Table 1.

$\beta = EI/k$	V1(r$_1$)	V1(r$_2$)	V1(r$_3$)	V1(r$_4$)	V1(r$_5$)	V1(r$_6$)	V1(r$_7$)
1	99.464	1.27	−0.854	0.19	−0.039	0.0076	−0.001
1000	Same values						
$\beta = EI/k$	V2(r$_1$)	V2(r$_2$)	V2(r$_3$)	V2(r$_4$)	V2(r$_5$)	V2(r$_6$)	V2(r$_7$)
1	1.292	96.624	3.173	−1.278	0.277	−0.054	0.0076
1000	Same values						
$\beta = EI/k$	V3(r$_1$)	V3(r$_2$)	V3(r$_3$)	V3(r$_4$)	V3(r$_5$)	V3(r$_6$)	V3(r$_7$)
1	−0.995	3.088	95.285	3.449	−1.335	0.277	−0.039
1000	Same values						
$\beta = EI/k$	V4(r$_1$)	V4(r$_2$)	V4(r$_3$)	V4(r$_4$)	V4(r$_5$)	V4(r$_6$)	V4(r$_7$)
1	0.0048	−1.402	3.393	95.237	3.455	−1.278	0.19
1000	Same values						

4.2 Cutting and Bending Stress Diagrams

To determine the expressions for shear forces *T(x)* and bending moments *M(x),* the functions are written using the step function Φ(x − a) in MATHCAD as follows:

$$\begin{aligned} T(x) := & -P1\cdot\Phi(x-d1) + V1\cdot\Phi(x-r1) + V2\cdot\Phi(x-r2) + V3\cdot\Phi(x-r3)\ \ldots \\ & + V4\cdot\Phi(x-r4) + V5\cdot\Phi(x-r5) + V6\cdot\Phi(x-r6) + V7\cdot\Phi(x-r7) \end{aligned} \tag{9}$$

$$\begin{aligned} M(x) := & -P1\cdot\Phi(x-d1)\cdot(x-d1) + V1\cdot\Phi(x-r1)\cdot(x-r1) + V2\cdot\Phi(x-r2)\cdot(x-r2) + V3\cdot\Phi(x-r3)\cdot(x-r3)\ldots \\ & + V4\cdot\Phi(x-r4)\cdot(x-r4) + [V5\cdot\Phi(x-r5)(x-r5) + V6\cdot\Phi(x-r6)(x-r6) + V7\cdot\Phi(x-r7)(x-r7)] \end{aligned} \tag{10}$$

The variation diagrams in Fig. 5 were obtained.

4.3 Diagrams of Displacements (Arrows) w (x) and Rotations fi (x)

To determine the expressions for displacements (arrows) w(x) and rotations fi(x), the bending moment integrals written using the step function Φ(x − a) in MATHCAD as follows:

$$\begin{aligned} FI(x) := & P1\cdot\Phi(x-d1)\cdot\frac{(x-d1)^2}{2} - V1\cdot\Phi(x-r1)\cdot\frac{(x-r1)^2}{2} - V2\cdot\Phi(x-r2)\cdot\frac{(x-r2)^2}{2} - V3\cdot\Phi(x-r3)\cdot\frac{(x-r3)^2}{2}\ldots \\ & + V4\cdot\Phi(x-r4)\cdot\frac{(x-r4)^3}{6} + V5\cdot\Phi(x-r5)\frac{(x-r5)^3}{6} - V6\cdot\Phi(x-r6)\frac{(x-r6)^3}{6} - V7\cdot\Phi(x-r7)\frac{(x-r7)^3}{6} \end{aligned}$$

$$W(x) := P1\cdot\Phi(x-d1)\cdot\frac{(x-d1)^3}{6} - V1\cdot\Phi(x-r1)\cdot\frac{(x-r1)^3}{6} - V2\cdot\Phi(x-r2)\cdot\frac{(x-r2)^3}{6} - V3\cdot\Phi(x-r3)\cdot\frac{(x-r3)^3}{6} \ldots$$
$$+ V4\cdot\Phi(x-r4)\cdot\frac{(x-r4)^3}{6} + V5\cdot\Phi(x-r5)\cdot\frac{(x-r5)^3}{6} - V6\cdot\Phi(x-r6)\cdot\frac{(x-r6)^3}{6} - V7\cdot\Phi(x-r7)\cdot\frac{(x-r7)^3}{6}$$

The variation diagrams in Fig. 6 were obtained.

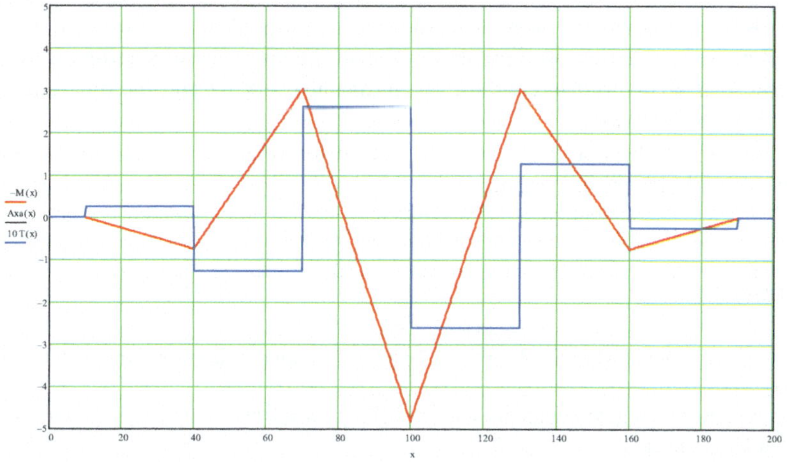

Fig. 5. Variation diagrams of shear forces T (x) and bending forces M (x) (β = 100)

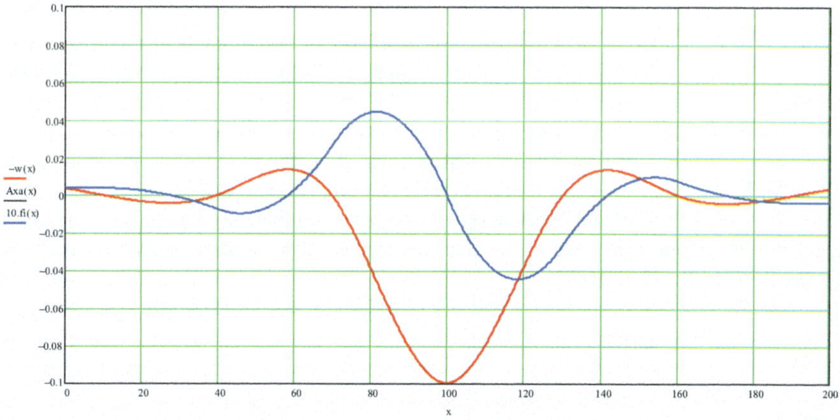

Fig. 6. Variation diagrams of displacements w (x) and rotations of section fi (x) (β = 100)

5 Conclusions

The continuous beams studied in this paper are hyperstatic systems formed by a bar supported on seven elastic supports with proportional elastic behavior. Solving such problems is very difficult analytically. Using the matrix methods and the facilities of the MATHCAD professional software, the following problems can be solved.

- The functions of the reactions in the support of the distance y of the load train from the end of the bar can be expressed analytically;
- The functions of the sectional forces T (x) and M (x) for a given distance y = d1 can be expressed analytically and the variation diagrams of these efforts can be drawn;
- The functions of the displacements w (x) and the rotations of sections fi (x) can be expressed analytically for a certain given distance: y = d1 and the diagrams of their variation can be drawn;
- Simulations can be made on the influence of the relative stiffness of the deformations of the hyperstatic system.

References

1. Marin, C.: Rezistenţa Materialelor Partea I - Solicitări simple, Editura Bibliotheca Târgovişte (2013)
2. Marin, C.: Rezistenţa Materialelor Partea a IIa - Teoria elasticităţii şi solicitări complexe, Editura Bibliotheca Târgovişte (2014)
3. Marin, C.: Probleme Tip De Rezistenţa Materialelor Rezolvate în Mathcad, Editura Bibliotheca Târgovişte (2012)
4. Marin, C.: The simulation of a train moving loads on a continuous beam with seven rigid supports located at the same level (i). Romanian Rev. Precis. Mech. Opt. Mechatron. **46**, 55–60 (2014)
5. Marin, C.: The simulation of a train moving loads on a continuous beam with seven elastic supports located at the same level (ii). Romanian Rev. Precis. Mech. Opt. Mechatron. **46**, 61–66 (2014)
6. Marin, C.: Matricial method used for reaction forces computation in the case of continuous beams supported by stiff bearings (i, ii). Romanian Rev. Precis. Mech. Opt. Mechatron. **45**, 142–145 (2014)
7. Marin, C., Ene, G.: Mathcad application to design of the discrete contact loose riding rings included in the supporting systems of rotary drums. Sci. Bull. Valahia Univ. – Mater. Mech. - SBMM **5**(8), 167–172 (2010)

STOP Sign Detection Using Python Programming

Sîrbu Cătălina, Macovei Dragoş, Rusu Dan Andrei, Grigore Alexandru, and Bogdan Grămescu(✉)

University Politehnica of Bucharest, 313, Splaiul Independenţei, 060042 Bucharest, Romania
bogdan.gramescu@upb.ro

Abstract. Autonomous driving systems are constantly evolving, in order to increase safety. One of the key elements is the recognition of traffic signs. The paper presents a solution in which a small scale demonstrator car is able to recognize the stop signs met on the road using Python libraries like OpenCV and NumPy in order to perform colors operations.

Keywords: Autonomous driving · Computer Vision · Traffic sign

1 Introduction

The autonomous car could be defined like a vehicle able of sensing its environment and operating without human involvement. A human passenger can is not required to take control of the vehicle at any time. He is not required to be present in the vehicle at all. An autonomous vehicle is able to go anywhere a traditional car goes [1].

The theme of the Bosch Future Mobility Challenge [2] proposes exactly this aspect only on a car (see Fig. 1) at a smaller scale (1:10). The ideal of the competition is the implementation of ideas that mimic real traffic situations like maintaining between lanes, parallel parking, recognition of pedestrians and traffic signs, bypassing obstacles, and reaction to traffic lights. The car has the following components: Raspberry Pi 4 board [3], Nucleo F401RE controller [4], Pi Camera [3], motor driver, LiPo Battery, servomotor, housing, chassis, and DC/DC converters for supplying the components.

Further, we will approach the image recognition process for the stop sign. The rest of the commands have already been implemented.

Fig. 1. Small scale demonstrator car (1:10), with housing (left) and without housing (right).

G. I. Gheorghe (Ed.): ICOMECYME 2020, LNNS 143, pp. 215–220, 2020.
https://doi.org/10.1007/978-3-030-53973-3_23

2 Fields of Use

Considering the high number of cars on the street, different problems appeared, such as decreasing the level of walkability and livability, traffic congestion, transportation costs (fuel, infrastructure), crowded or insufficient parking spaces, increasing urban CO_2 emissions.

In order to reduce and eliminate this type of problem, in the automotive industry have been created several levels of driving automation from Level 0 - No automation (Manual Control) to Level 5 - Full automation (No human interaction required). At this moment, fully autonomous cars are not available to the general public, but they are continuously tested in specialized centers.

Increasing the safety of autonomous cars depends on the level of intelligence that these mechatronic systems integrate. A key element is the recognition of traffic signs.

3 Analysis of Existing Solutions

The digital branch of photography became commercially available in 1990. Later, the issue of how computers can gain information from digital images or videos arose.

There are various online applications with which it is possible to detect objects, shapes, colors, textures, whether it is day or night and so on. Some of those applications are based on concepts such as Computer Vision [5], Machine Learning and Artificial Intelligence.

In a real case in which the driver meets the stop sign, he processes the information, realizes the meaning of the traffic sign, and makes a decision. The goal of computer vision is to succeed in accomplishing the same task at the same time or faster. These algorithms are implemented on computers.

In addition to the software part, the hardware part is also required. A real-life autonomous car needs a radar, a lidar, several cameras but the competition car consists of a motor controller, an image processing controller, a compatible camera (see Fig. 2).

Fig. 2. Hardware required: Nucleo F401RE (left), Raspberry Pi 4 model B (middle), Pi Camera (right)

4 Own Solution

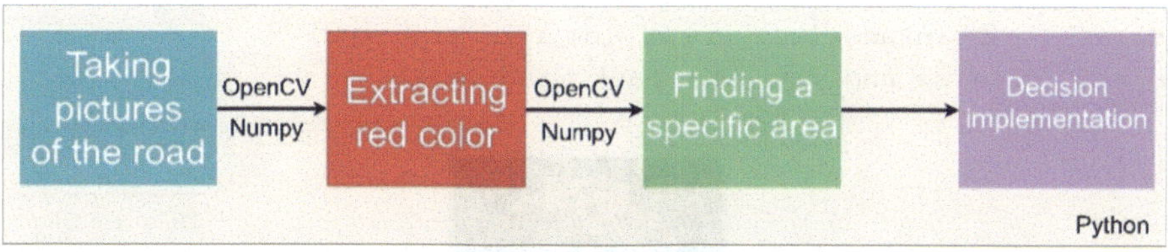

Fig. 3. Block diagram

Video processing is performed using the NumPy (1.18.1 version) and OpenCV (4.2.0 version) software modules, these being programmed in Python (3.8.1 version) (Fig. 3).

```
import cv2
import numpy as np
```

To get started, this program uses the Raspberry Pi camera to take pictures of the route in real time. To recognize the red color, we need a range of BGR colors (minimum and maximum values) because the shade of red may differ depending on the light intensity. We used a special function to transform the color space from BGR to HSV. Thus, we positioned the sign at different angles to see all the possibilities of values for the interval. This implementation gives us a wider range for sign recognition. An example of a group of BGR values can be seen in Fig. 4. To exemplify the implementation, we used a sample image with the stop sign on the road, more precisely from the intersection.

```
hsv_frame = cv2.cvtColor(img, cv2.COLOR_BGR2HSV)
low_red = np.array([76, 53, 228])
high_red = np.array([114, 92, 255])
```

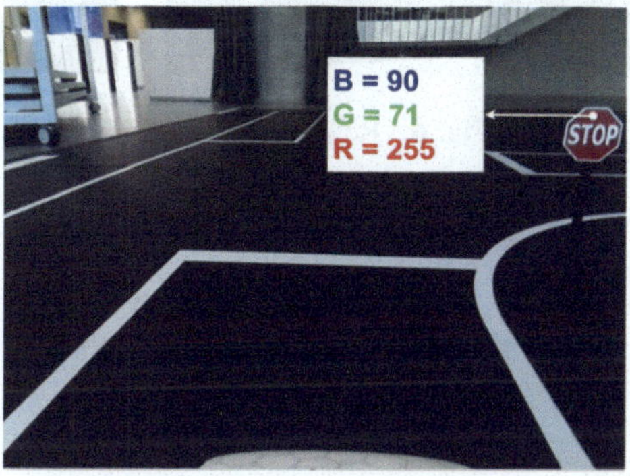

Fig. 4. BGR values for red

The next step is to add a black mask over the original image to highlight only the color range to be processed (see Fig. 5).

```
red_mask = cv2.inRange(hsv_frame, low_red, high_red)
red = cv2.bitwise_and(img, img, mask = red_mask)
```

Fig. 5. Black mask

Given the fact that we can meet other red objects that fall within the chosen range, we have implemented a region of interest (ROI) (see Fig. 5) on the right side of the image because in the case of the contest only there can be the stop sign. For this implementation, the width and length of the image must be known. For sketching the ROI, the upper left corner is represented by the coordinates (x1, y1) and the lower right corner is represented by the coordinates (x2, y2). In order to be able to use these coordinates, they must be placed in a certain order as follows:

```
ROI = img[y1:y2, x1:x2]
```

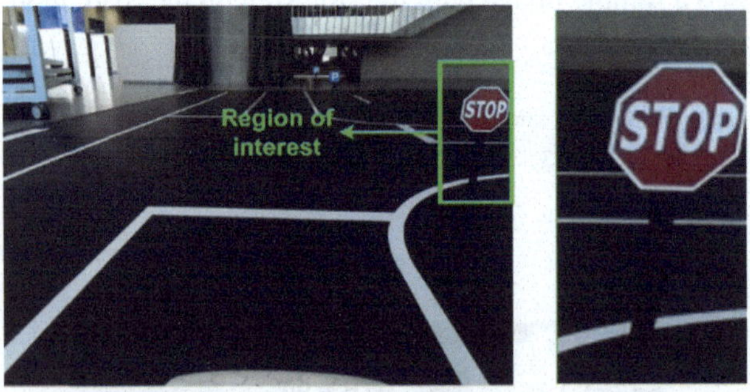

Fig. 6. Region of interest

Another precaution to avoid confusion with other objects of the same color is to check the area of our color. The stop sign is a fixed object that does not change its size, so its area will remain the same in most cases, obviously excluding situations in which it is slightly inclined. To calculate the area, we first need the contour of the object. To do this, *moments* is used because it skips searching in the contours array, computes the area based on moments and it is much faster.

```
Contours = cv2.findContours(filter, cv2.RETR_EXTERNAL,
cv2.CHAIN_APPROX_SIMPLE) ...
moments = cv2.moments(contours[0], True)
area = moments["m00"]
```

Given that the image is dynamic, it is necessary to have a working interval because the fixed values are not suitable for any situation encountered on the route. As can be seen in Fig. 6, zone 1 represents a red area with a value that falls within the chosen range and in zone 2 there is a much smaller area than the ideal one (Fig. 7).

```
if lowerAreaThreshold < area < upperAreaThreshold:
print("Area checked")
```

Fig. 7. Ideal (1) and small (2) red areas

When the object encountered has a certain color, a certain area and it is in the region of interest chosen earlier, it means that we have met the stop sign and further we have to give the command to the engine to make the decision to stop.

All this processing is possible thanks to the Raspberry Pi 4, a low cost, credit card sized computer, that plugs into a computer monitor or TV, and uses a standard keyboard and mouse. This little device enables people to explore computing, and to learn how to program in languages like Scratch and Python. It is capable of doing everything you'd expect a desktop computer to do, from browsing the internet and playing high-definition video, to making spreadsheets, word-processing, and playing games [3].

Through serial communication the Raspberry PI sends the processed information to the Nucleo. This board is a microcontroller, and in this situation a personal library is used. With its help you can change the speed and angle of the wheels through the servomotor. It also sends the command to the motor via the PWM signal and a motor driver containing the H-bridge.

```
import SerialHandler
SerialHandler.driveWithConstPower(2000)
SerialHandler.setAngle(f.normalize(float(speed)))
```

5 Conclusion

Although this project took place around a small-scale demonstrator car, the ideas can be implemented and used in real life.

In carrying out this project, all the measures that make the working method and material resources more efficient were taken into account.

Even if only the recognition of the stop sign was presented, the algorithm can be implemented for the recognition of parking signs, pedestrian crossings, priority roads and much more.

References

1. Synopsys Page. https://www.synopsys.com/automotive/what-is-autonomous-car.html. Accessed 28 May 2020
2. Bosch Future Mobility Challenge Homepage. https://www.boschfuturemobility.com/. Accessed 28 May 2020
3. Raspberry PI Product Page. https://www.raspberrypi.org/. Accessed 28 May 2020
4. STM32 Product Page. https://www.st.com/content/st_com/en.html. Accessed 28 May 2020
5. Computer Vision Wikipedia Page. https://en.wikipedia.org/wiki/Computer_vision. Accessed 28 May 2020

Upgrading Obsolete Hydraulic Power Units to Become Remotely Monitored, Energy Efficient and Intelligent

Mihai Avram and Valerian-Emanuel Sarbu(✉)

POLITEHNICA University of Bucharest,
313 Spl. Independentei, Bucharest, Romania
vale_sarbu@yahoo.com

Abstract. The hydraulics domain is experiencing a fast progress toward intelligent, energy efficient, IoT and edge connected devices. The 4th industrial revolution (I4.0) expanded greatly since it started but now there are many devices left behind that are either too expensive to upgrade or the manufacturer offers no support for upgrade options. Therefore, as a proof of concept, a fixed flow Rexroth Hydraulic Power Unit was upgraded to be smart and efficient using custom made smart devices providing control and IoT Gateway functionality. The resulting system is more efficient, quieter, has lower operating costs and can be remotely monitored.

Keywords: IoT · Smart · Hydraulics · Remote management

1 Introduction

Ever since the first industrial revolution took place a trend to automate the machines, reduce the maintenance cost and increase the efficiency has been ongoing for the industrial manufacturing business. Now more than ever with concerns over pollution, strict regulations from countries and the new social distancing this trend is moving at a very fast pace. Remote managed, intelligent, edge controlled, and highly efficient devices are seeing a very high demand. Given these trends, obsolete equipment with lower efficiency that also need human supervision are just thrown out and replaced; this is the case for hydraulic power units, especially the ones capable of fixed flow rate as they have a low initial cost and very high operating cost [1, 2].

Hydraulic systems are known to be a good energy source when high power and torque is desired while being compact and low maintenance. The hydraulic medium is also the cooling agent and lubricant for the system further simplifying the system construction.

Advances in pump manufacturing techniques enables higher efficiency through variable flow rates obtained by pump construction or by varying the input shaft speed; some use high power DC motors or even brushless motors that require power electronics but offer even better control and efficiency. Most of the older models however use Induction motors that were just connected to the grid with no speed control possible.

G. I. Gheorghe (Ed.): ICOMECYME 2020, LNNS 143, pp. 221–230, 2020.
https://doi.org/10.1007/978-3-030-53973-3_24

This means that the output flow is always the maximum hydraulic pump can deliver and when the full flow is not needed the excess is converted to heat/losses [3, 4].

Figure 1 shows an example of different energy cost comparisons depending on the technology used. Fixed pumps are obviously the worst when it comes to efficiency as seen in the graph and variable speed pumps with intelligent electronic drivers are the best.

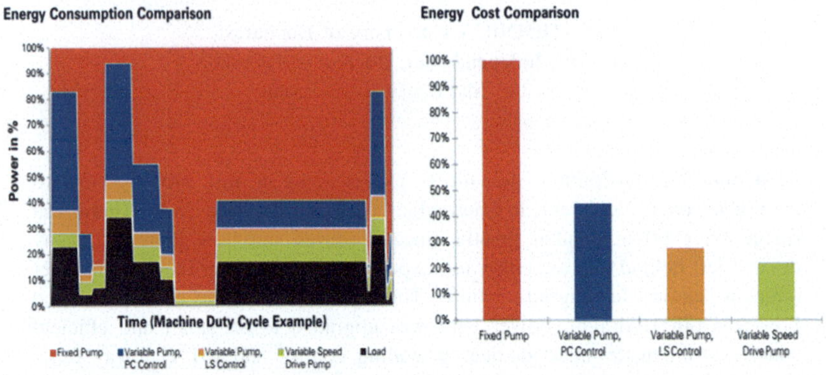

Fig. 1. Comparison of different pump configurations [12]

The result of this high energy consumption makes the operating cost exceed greatly the initial purchase cost [5, 6].

Energy is however not the only cost these types of machines incur during their lifetimes. The need for an operator or technician to operate and perform maintenance also adds to this cost. Unforeseen malfunctions are often caused by user error (lack of maintenance) and are very expensive in the industry as they cause a complete halt of the production line until the fault is fixed. An intelligent, self-diagnosing and remotely monitored system can pay for itself in a relatively short amount of time [7, 8].

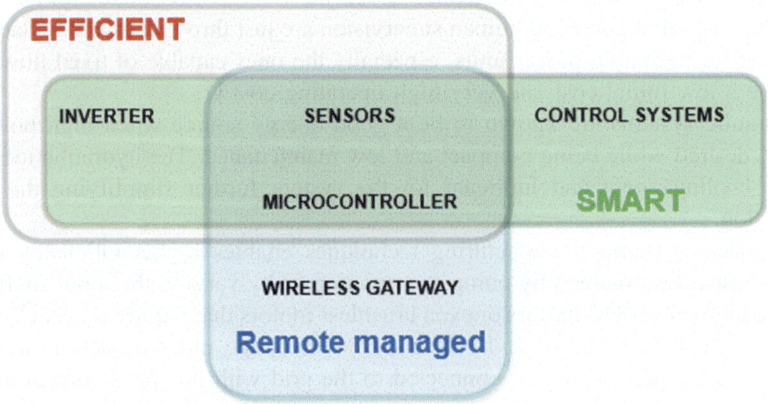

Fig. 2. Overlap of requirements

Therefore, this paper aims to study the upgrade potential of a hydraulic power unit using low-cost locally available hardware (where possible) and bring new life into an obsolete unit.

In order to resolve the previously stated problems the following equipment are necessary, as also visible in Fig. 2:

- Inverter – to improve efficiency and add possibility of diagnostics.
- Sensors – to provide feedback about the current state of hydraulic circuit and environment.
- Control board – to interface to the inverter, sensors, actuators and provide safety features.
- IoT Gateway – to allow remote diagnostics and in the future, management.

2 Hardware Setup

The original setup is displayed in Fig. 3 and portrays a standard power unit from Rexroth, namely ABSKG-60AL9/VGF2-016/112 M-4-B, along its core schematic. This schematic does not include the auxiliary components as: Analogue pressure indicators, filter block and cooler since they won't be altered.

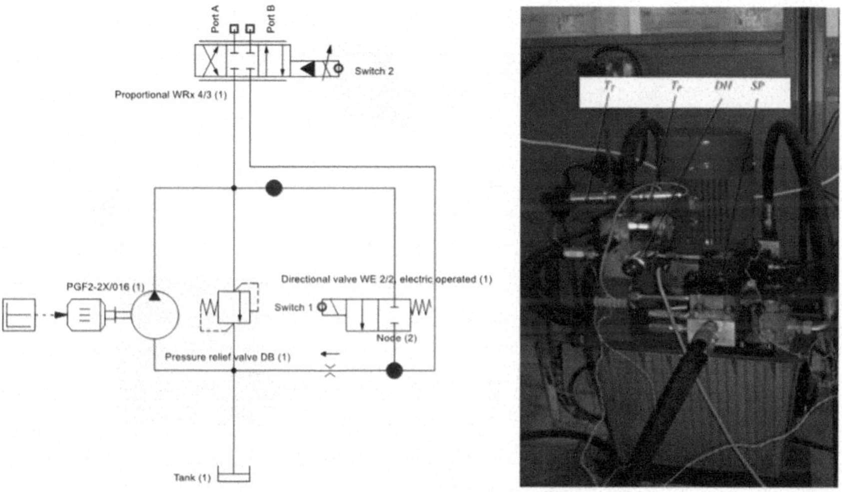

Fig. 3. Original setup with schematic [13]

This basic setup provides a constant flow of 24 L/min at a rotational speed of 1500 RPM to the downstream system at the pressure limit set by the pressure relief valve or below it. A bypass valve can be used to "short" the flow directly to tank, limiting the pressure to almost 0 to reduce power draw. Depending on the downstream system, this type of system is almost guaranteed to be inefficient since it can only provide full flow or no flow. The proportional valve with an amplifier is used to limit the flow by varying the opening of an orifice. Figure 4 illustrates a thermal image capture of the whole pump and the flow control valve, losses are clearly seen as heat, underestimated due to the cooling power of the hydraulic medium [9].

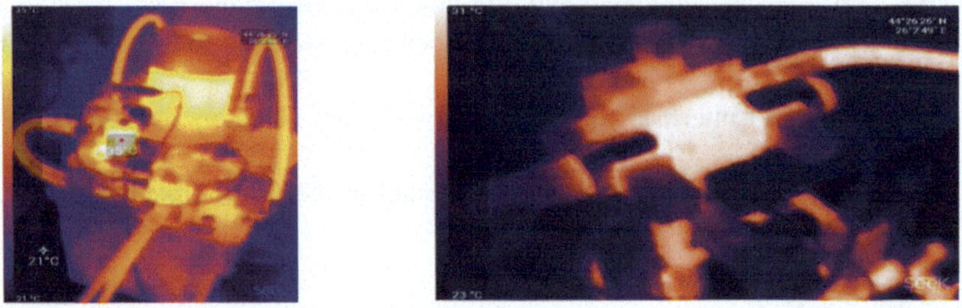

Fig. 4. Thermal camera capture of power unit and flow control valve

As mentioned before the new system needs to incorporate more sensors, an inverter, control system and an IoT gateway in order to be qualified as smart, efficient and remotely managed.

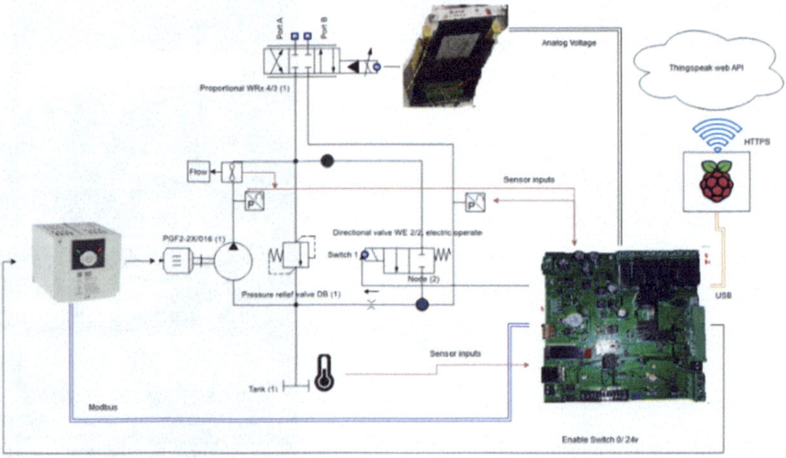

Fig. 5. Final design

As seen in Fig. 5, the original structure has been altered to contain the following:

- 2 pressure sensors, one on the output of the pump for pressure regulation and one on the return path for filter and heat exchanger monitoring, clogging will cause excessive return path pressure.

- One flow rate sensor, mandatory for matching the flow rate needed in the system (control loop) and for detecting faults.
- Temperature sensors for monitoring the hydraulic fluid temperature, and a low resolution/low-cost thermal camera for simultaneous multipoint measurement of the power unit.
- Proportional electrically controlled pressure relief valve for fast pressure regulation.
- Proportional 4/3 valve with spool position controller, voltage controlled
- IoT Gateway – raspberry pi
- Control board – custom made for hydraulics.

Since most of the setup used standard available components a short description for the control board is necessary (Fig. 6):

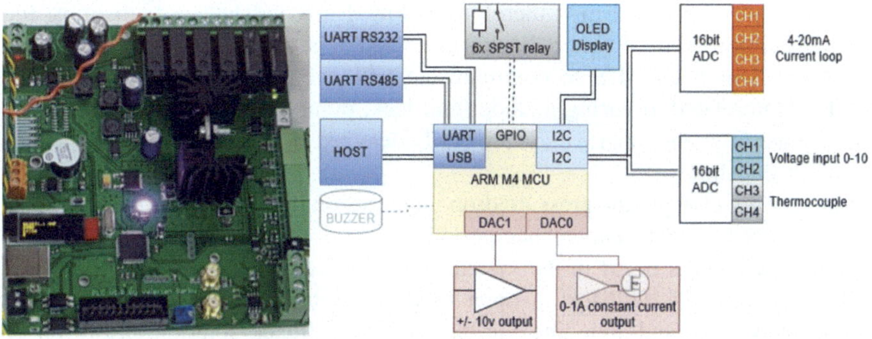

Fig. 6. Control board and block diagram

The control board was built around the new ATSAML21 microcontroller from Microchip. It was created as an all-in-one solution for energy efficiency by incorporating both voltage output (for existing hydraulic controllers) and constant current 0–1 A for controlling proportional equipment. It also contains multiple relays for other switchable valves or enable signals, RS485 or RS232 for interfacing to other equipment (Inverter in this case), USB for interfacing with optional host or IoT gateway and finally two digital optically isolated inputs for level oil level switches that are usually present on the tank.

The IoT gateway task is attributed to a Raspberry PI board. This was done because MATLAB provides very good support for this board and it is the "Go to" for developing smart solutions. They also have an Internet of Things platform called ThingSpeak. This platform is free with certain limitations that are suitable for entry-level remote management [10].

3 Software Implementation

The software was designed to firstly regulate pressure at a set value and secondly to adjust the flow in order to minimize losses.

Pressure regulation: the constant current sink regulator of the control board is used to adjust the electronically controlled pressure relief valve by constantly adjusting the opening of the valve in order to maintain set pressure.

Flow control: in order to optimize energy usage, the previously mentioned relief valve must be almost closed (no flow returning to tank). This can be done in two ways, that function alternatively if one fails:

Measuring flow that goes to the system; by knowing the amount of flow the pump generates at each frequency and adjusting the frequency to a value a bit higher than is currently measured (to allow for pressure drops).

If the primary loop is disturbed (for example faulty flow sensor), the second loop can take over and use the command value for the proportional valve, trying to keep the valve opening as low as possible without knowing the actual flow rate.

For fault diagnostics the current loop sensors have an inherent diagnostics system as a value of 0 is in fact 4 mA; therefore, below 4 mA indicates a fault (most likely cable interruption). For inverter diagnostics RS485 is used as a hardware layer for either LSBUS or MODBUS to communicate with the inverter and check its status. Other faults unrelated to wiring include fluid leak, pump overheating, pump efficiency drop, clogged filter and others; these are easily detected by monitoring sensors against a baseline or thresholds.

As filters get clogged the pressure drop increases on the return line, the slope can be used to predict the next maintenance period.

As different faults occur in the system the software can also decide the next step: turn the machine off, fallback to open loop mode or continue while displaying a warning to the user and sounding the buzzer. The fallback solution allows for temporarily discarding energy savings for reliability, allowing the maintenance department to schedule the fix without disrupting production.

4 Results

In a first setup during the bring-up of the control board the pressure control loop was tested in order to verify that the system is functioning in parameters (Fig. 7).

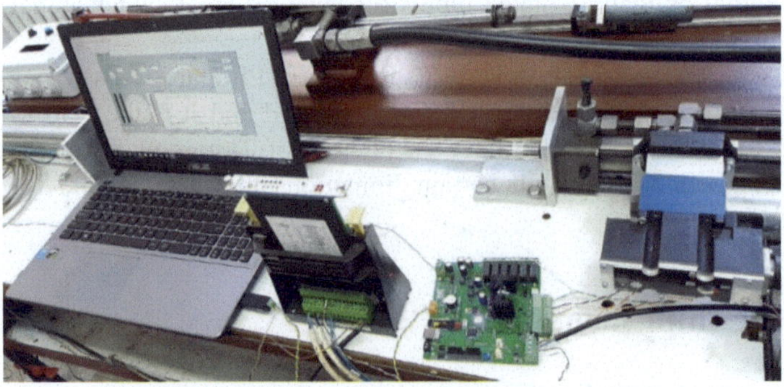

Fig. 7. First hardware bring-up with Labview

The control board was connected to a laptop running LabVIEW in order to graph the result with ease. The voltage output of the control board is attached to a proportional hydraulic valve controller board; it has the role of simulating a varying resistive load.

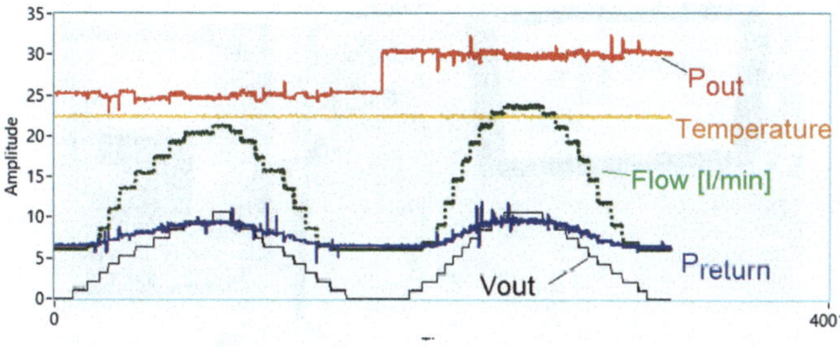

Fig. 8. Pressure regulation

Figure 8 contains many elements so in order to understand it first we must observe Vout as it is the proportional valve control signal. It is tracking an elliptical curve and as it traces it the flow and return pressure follows. This pattern is repeated twice at two pressure set points. The red trace Pout is the resulting pressure from the automatic pressure regulator loop. This figure clearly demonstrates the capability of the designed loop to maintain the set pressure with reasonable accuracy (at 25 and 30 bar).

For a better real-world test, the hydraulic power unit was connected to a system that positions a large hydraulic piston at 3 predefined locations (VT-HNC100) at maximum speed. This time the pressure regulator still worked though with spikes appearing at every full stop and start of the piston movement and a noticeable over lapped $\sim$3 bar ripple as it accelerates/decelerates (Fig. 9).

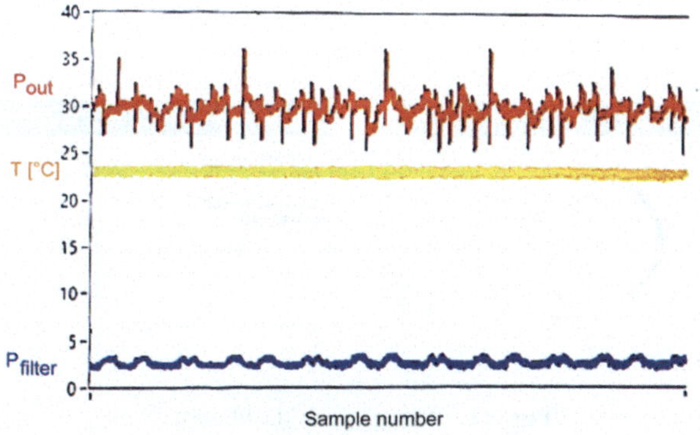

Fig. 9. Pressure regulation load step

The setup displayed in Fig. 10 was used to connect to ThingSpeak in order to send status updates. Updates are allowed every 15 s on the free version, but should be sufficient for an entry level I4.0 setup [10].

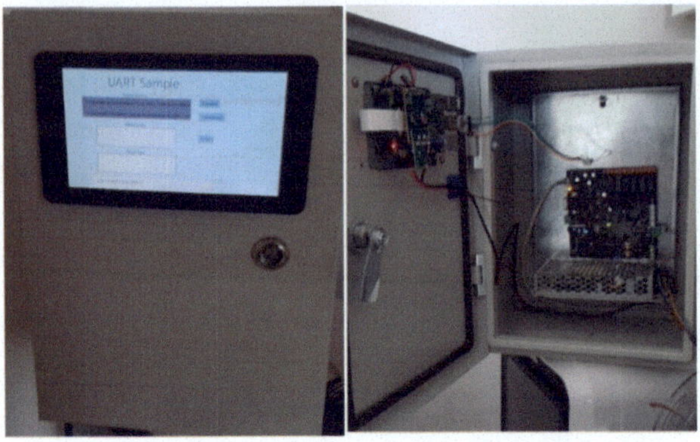

Fig. 10. Final setup with raspberry pi

The updates are easily made by just accessing a webpage with the API key and a field name equals value, example: status = All systems running

Figure 11 exhibits a snapshot from ThingSpeak webpage, with the updates recorded after a short time online.

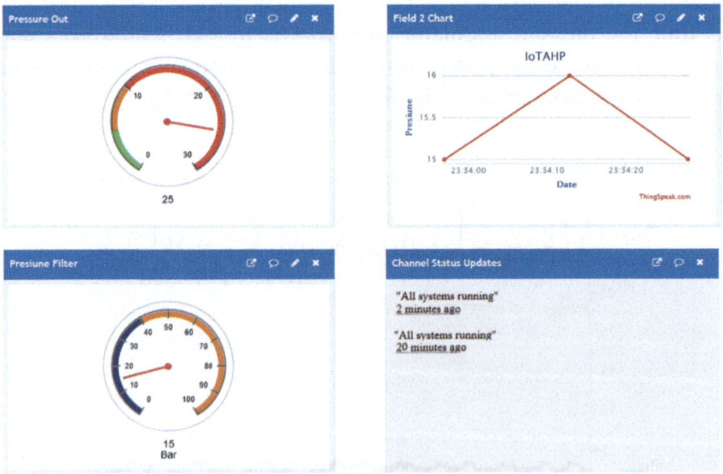

Fig. 11. ThingSpeak dashboard

5 Conclusions

With the recent advances in hydraulics many systems were left behind as being inefficient. Using simple readily available parts this paper can hopefully prove that any hydraulic power unit can be upgraded to become efficient, smart and cost saving. Reducing waste of any kind is very important in times like this as it not only reduces operating cost but also being responsible with waste management attracts investors. All points were not fully featured in this paper; the setup is however fully capable of accomplishing them as the components were carefully selected specifically for this purpose. The IoT provider used for remote management is lacking bidirectional support (remote control) but provided the fastest "time to market" possible. Microsoft Azure and Amazon AWS provide full services for bidirectional support and even AI capabilities that can further improve the efficiency and energy management, synchronizing maintenance, load sharing and so on [11].

References

1. Rexroth: Ready for Industry 4.0: Connected hydraulics, Rexroth, Iulie 2017. https://m.boschrexroth.com/en/web/xc/trends-and-topics/directions/ready-for-industry-4-0-connected-hydraulics. Accessed Ianuarie 2018
2. Martin, E.: Pump control: Which is the right one?, Rexroth, 29 Nov 2017. http://blogs.boschrexroth.com/en/topics/decision-making-hydraulic-pump-control/. Accessed 03 Ianuarie 2018
3. Gib, S.: Managing the risk of the Internet of Things, Control Engineering, 25 Aug 2015. https://www.controleng.com/single-article/managing-the-risk-of-the-internet-of-things/03e6419374662db00e02f0436cbd5cb2.html
4. Guide, T.I.: Hydraulics for Industry 4.0. instrumentation.co.za. http://www.instrumentation.co.za/55135n
5. Gunder, A.: Hydraulic power units 4.0. Bosch Rexroth, 22 Nov 2017. http://blogs.boschrexroth.com/en/pq-en/hydraulic-power-units-4-0/
6. Endres, M.: Pump control – simple or intelligent?, Rexroth, 6 Dec 2017. http://blogs.boschrexroth.com/en/topics/hydraulic-pump-control-systems-simple-or-intelligent/. Accessed 01 Ianuarie 2018
7. Electromechanical Team: Is Industry 4.0 Driving the Need for Smarter Motion Control Products?. Parker, Aug 2017. http://blog.parker.com/is-industry-40-driving-the-need-for-smarter-motion-control-products
8. Industry 4.0/ IoT – Products and solutions. Festo. https://www.festo.com/cms/nl-be_be/56644_56690.htm
9. Rexroth, B.: Datasheet PGF2x. https://md.boschrexroth.com/modules/BRMV2PDFDownload-internet.dll/re10213_2015-05.pdf?db=brmv2&lvid=1188621&mvid=13760&clid=20&sid=FA0A4812C9BBE797FCD1096A16DAFABA.borex-tc&sch=M&id=13760,20,1188621. Accessed 02 Jan 2019
10. MathWorks: "ThingSpeak," MathWorks. https://thingspeak.com/prices/thingspeak_standard. Accessed 27 May 2020

11. "AWS IoT Core," Amazon. https://aws.amazon.com/iot-core/
12. https://www.eaton.com/ecm/groups/public/@pub/@eaton/@hyd/documents/content/pct_1233971.pdf
13. Avram, M., Bucşan, C., Miu, S., Tanase, M.: Mechatronic Design of a Hydraulic Power-Supply Unit (2012)

Theoretical Analysis of a Hydraulic Energy Generation System Equipped with a Gear Pump

Mihai Avram, Valerian Sârbu, Emil Ionuț Niță(✉), and Lucian Bogatu

University "Politehnica" of Bucharest, 060042 Bucharest, Romania
nitaemilionut@yahoo.com

Abstract. This paper presents the theoretical study of a hydraulic power unit that uses an external gear pump. For this power unit, a schematic is firstly brought up, the working parameters are enumerated and finally its mathematical model is produced. The resulting model is adapted for numerical simulation and simulated using two software platforms: MATLAB Simulink and Simcenter Amesim. The results are later compared.

Keywords: Simulation · Hydraulics · Pumps

1 Introduction

A first step in designing a theoretical analysis of a hydraulic system is the development of the mathematical model of the desired system. This mathematical model may be considered as an abstract system, equivalent in some aspects with the real system but easier to model and compute; the information gained from the study of the mathematical model defines, with a certain degree of precision, the behavior of the physical system [1–4].

The study of real fluid flow is generally difficult to approach by complete theoretical methods, mainly due to the special complications caused by a series of disturbing factors such as fluid friction (viscosity), turbulence, occurrence of solid particles, the presence of internal heat sources, etc. For this reason, non-essential disruptive factors are frequently eliminated and only those with a significant effect are retained to model the system. A simplified theoretical model is thus obtained that exhibits sufficient accuracy and may allow describing the most important processes of the real fluid dynamics [5].

In hydraulic systems, the fluids used to make the equipment work are an important transport medium for energy and information [6, 7]. Mineral or synthetic oil that contains specific additives used to increase the systems performance or to adapt the fluid to the environmental conditions are preferred in the hydraulics systems. Therefore, the fluid is an important aspect of a mathematical hydraulic system model.

It is well known that in a hydraulic system the flow can be adjusted using two methods: volumetric and resistive [8, 9]. The volumetric method consists in changing the pump flowrate, at variable pressure, depending on the load driven by the motor, while the resistive method is achieved by varying the local resistance of the

G. I. Gheorghe (Ed.): ICOMECYME 2020, LNNS 143, pp. 231–241, 2020.
https://doi.org/10.1007/978-3-030-53973-3_25

supply/exhaust pipe connected to the motor, at constant pressure, using a variable hydraulic resistance. Usually, volumetric method is used where high powers are controlled. However, if a fast and precise power transfer is required, up to a limit of 10 kW, then the resistive method must be used with the compromise of a lower efficiency (consequence of a local pressure loss on the adjusted section and a discharge to the tank through the safety valve of excess flow). In order to minimize power losses and increase dynamic performance there is the possibility of combining the two control methods.

The pump chosen to generate hydraulic energy incurs a significant share of the total hydraulic drive system cost. If it is of fixed flow rate type, the price of the system is much lower however the energy losses it suffers during operation are much more significant. Many users do not consider this important aspect and are opting for a fixed flow rate pump. One of the most purchased hydraulic pumps is the one with external gears.

The pump is integrated in a system, called hydraulic power generation and distribution system or more briefly hydraulic power unit [8, 9]. This usually contains: a pump, a tank, a safety valve, a heat exchanger, filters, etc. Its role is to ensure that the drive station that it is designed for gets the required oil flow rate under a certain working pressure. At the same time, it must maintain the temperature of the fluid within normal limits and ensure its purity (required fines filtration).

Flow adjustment can be done by volumetric method, changing the pump displacement or the drive shaft speed [10, 11]. Geared pumps with external gears have a fixed displacement although using a motor with a variable speed controller the flow rate that the pump provide the to the system can be changed. This is the type of structure that this paper analyzes.

The static and dynamic performances of a hydraulic pumping system can be determined in a theoretical or experimental manner.

Theoretical approach is preferred in one of the following situations:

- in system designing stage where the aim is to identify the most suitable components that meets the requirements of the designing theme, as well as to determine the static and dynamic performances of the future system;
- when the system exists physically and the aim is to optimize its functional parameters.

Almost in all situation, the theoretical analysis should precede the experimental testing of the system; In this way the optimal test conditions can be determined taking into account the fact that the parameters defining the system can be changed very easily. Moreover, test conditions that are not always possible to perform and repeat in the laboratory can be considered in the theoretical analysis.

The theoretical study of the system requires the following steps:

- realizing the constructive - functional analysis of the equipment; this activity must be completed with a schematic diagram where both the constructive and the functional parameters are mentioned; this schematic diagram must also highlight the operation principle of system;

- establishing the most important variables in the system; this selection is made according to the schematic diagram and it concludes with a table of variables;
- establishing the simplifying hypotheses; adopting ideal conditions for the least significant parameters in the system
- establishing the mathematical model of the pumping system; it contains a number of equations that describe the functioning of system's components; these equations are according to the mechanical and fluid universal laws (conservation of energy, conservation of mass, conservation of momentum, etc.);
- geometric and structural optimization of the system.

Following the steps mentioned above results a mathematical model which contains a number of differential equations representing the behavior of the physical system. In some cases, such a high complex model cannot be used directly in computer simulation, requiring a simplification of the model that should not change the system behavior.

2 Schematic Diagram of the Hydraulic Energy Generation System

Figure 1 presents the schematic diagram of the hydraulic system, containing the following parts: gear pump (P), tank filter (F_t), safety valve (Ssig), hydraulic tank (Rz), three phases electrical motor (M), rectifier (R), electrical filter (F), inverter (I) and the frequency command (f). Figure 2 presents the schematic structural diagram for geometric and constructive parameters, specified on Table 1.

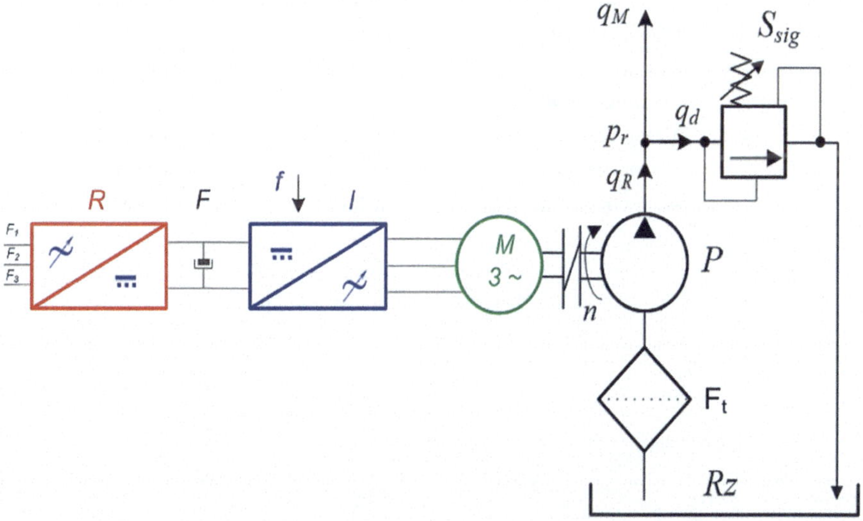

Fig. 1. Schematic diagram of the hydraulic energy generation system

Table 1. Geometric and constructive parameters of the system

Parameter	Name	Symbol, measure unit	Value/Variation range
Functional parameter	Exhaust flow rate of the pump	$q_r[l/min]$	–
	Motor flow rate	$q_M[m^3/s]$	$1.15 \cdot 10^4$
	Flow rate returned to the tank	$q_d[l/min]$	–
	Working pressure	$P[N/m^2]$	$40 \cdot 10^5 \ldots 100 \cdot 10^5$
	Atmospheric pressure	$P_0[N/m^2]$	$1,013 \cdot 10^5$
Fluid parameters (H46EP)	Modulus of elasticity	$E\ [N/m^2]$	$1,6 \cdot 10^9$
	Fluid density	$\rho_{15^oC}\ [Kg/m^3]$	895
	Dynamic viscosity	$\eta\ [Ns/m^2]$	0, 04896
Structural gear pump parameters	Number of gear teeth	$z\ [-]$	10
	Gear module	$m\ [mm]$	3
	Coefficient of profile shift	$x\ [-]$	0, 3
	Gear width	$b\ [mm]$	10
	Normal pressure angle	$\alpha_0\ [^o]$	20^0
	Radius of pitch diameter	$r\ [mm]$	15
	Radius of base diameter	$r_b\ [mm]$	14, 095
	Radius of outside diameter	$r_a\ [mm]$	18, 90
	Radius of root diameter	$r_f\ [mm]$	12, 15
Structural valve parameters	Valve angle	$\alpha\ [^o]$	15
	Valve diameter	$d\ [mm]$	8
	Piston diameter	$D\ [mm]$	16
	Piston width	$L\ [mm]$	20
	Piston clearance	$j\ [mm]$	$8 \cdot 10^{-3}$
	Valve mass	$m\ [Kg]$	0, 04
	Spring stiffness	$k_{arc}\ [N/m]$	30000
	Reference volume	$V\ [m^3]$	$2 \cdot 10^{-6} \ldots 5 \cdot 10^{-6}$

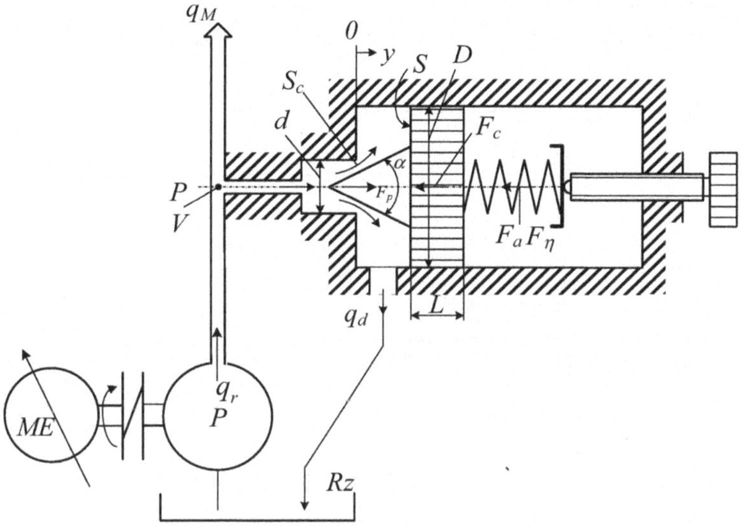

Fig. 2. Schematic structural diagram of the system

3 Mathematical Model of the System

The aim of this paper is to develop a mathematical model as accurate as possible, avoiding the utilization of simplified hypotheses. The mathematical model was established correlated with Figs. 1, 2 and Table 1.

The mathematical model contains the following equations:

A. Equation of the average flow rate of the pump related to the rotation angle φ [12]:

$$q_r = b \cdot \left\{ k_1 - k_2 \cdot \left[\varphi - k_3 - \frac{\varphi - k_3}{\varphi_p} + 0.5 \cdot \varphi_p \right]^2 \right\} \cdot \omega \left[\frac{cm^3}{s} \right] \tag{1}$$

where:

- $k_1 = r_a^2 - r^2$
- $k_2 = r_b^2$
- $k_3 = \tan(\alpha_0)$
- $\varphi_p = (2 \cdot \pi)/z$

In relation (1), ω represents the angular velocity of the electrical motor shaft, expressed in [rad/s].

For the considered gear results:

- $k_1 = 132{,}21\ mm^2$
- $k_2 = 198{,}67\ mm^2$

- $b = 12$ [mm]
- $k_3 = 0{,}364$ [–]
- $\varphi_p = 0{,}6283$ [–]

The angular velocity equation of the motor shaft is:

$$\omega = \frac{2\pi}{60} \cdot \frac{f}{p} \cdot (1 - g) \tag{2}$$

where:

- f – command frequency in Hz
- p – number of motor's pole pairs
- g – slip of the motor

B. Equation of the flow section of the Sc valve:

$$S_c = k_{s1} \cdot y - k_{s2} \cdot y^2 \tag{3}$$

where:

- $k_{s1} = \pi \cdot \sin(\alpha) \cdot d$
- $k_{s2} = \pi/2 \cdot \sin(\alpha) \cdot \sin(2 \cdot \alpha)$
- y – valve position

C. Pressure differential equation of reference volume V:

$$\frac{dP}{dt} = \frac{E}{V} \cdot \left[q_r - q_M - A \cdot \frac{dy}{dt} - S_c \cdot \sqrt{\frac{2}{\rho} \cdot (P - P_0)} \right] \tag{4}$$

where:

- $A = \pi \cdot \frac{D^2}{4}$ [mm^2]

D. Valve motion equation:

$$m \cdot \frac{d^2y}{dt^2} = P \cdot A^* - F_{a0} - k_{arc} \cdot y - k_{f1}(P - P_0) \mp k_{f2} \cdot \frac{dy}{dt} - 2 \cdot (P - P_0) \cdot S_c \cdot \cos\alpha \tag{5}$$

where:

- $A^* = \frac{\pi d^2}{4}$
- $F_{a0} = A^* \cdot P_r$
- $k_{f1} = \frac{\pi \cdot D \cdot j}{2}$
- $k_{f2} = \frac{\pi \cdot D \cdot j}{L} \eta$

4 Simulation of the System

Equations (1)–(5) form the mathematical model of the proposed hydraulic power generation system. The simulation of this model is preceded by the establishment of the initial conditions, as follows:

$$\begin{cases} q_M = 1.15 \cdot 10^4 \frac{m^3}{s} \\ P = P_0 \\ y = 0 \\ \frac{dy}{dt} = 0 \\ S_c = 0 \end{cases} \quad (6)$$

Matlab Simulink and Amesim Simcenter graphical programming environments are used for the numerical integration of the mathematical model. Figure 3 represents the block diagram of the program realized in Matlab Simulink. The graphical blocks (Fig. 3) contain mathematical functions associated to the mathematical model of the pump, motion equation of the pressure valve and the fluid flow equation through the valve adjusting section. The program contains:

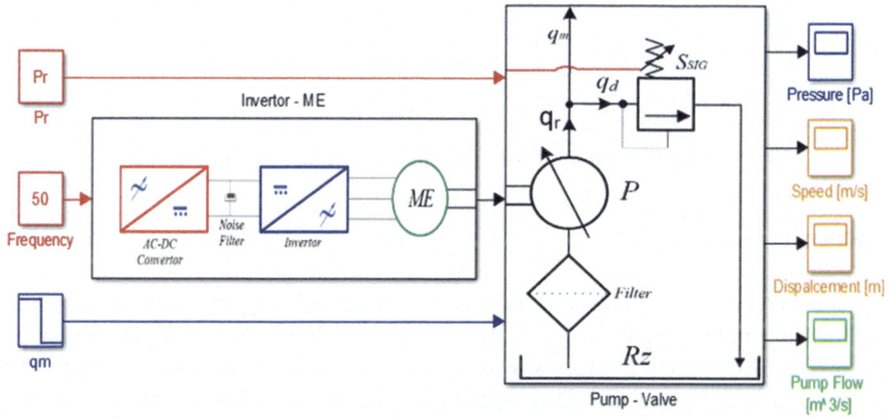

Fig. 3. Block diagram of the program realized in Matlab Simulink

- Input parameters: required pressure (P_r), motor flow rate (q_M) and the control frequency (f) for the inverter
- Motor Control Block
- Pump-Safety valve block
- Parameters of interest: system pressure, pump flow rate, valve speed and displacement

For validation of the results obtained from Matlab Simulink, another simulation model was developed in the Amesim Simcenter environment (Fig. 4). This model contains graphical blocks, each with a well-defined mathematical relationship. Adding them together results a virtual system as close as possible with the real one. An adjustable hydraulic resistance (working as system load; same as q_M) was introduced in the model.

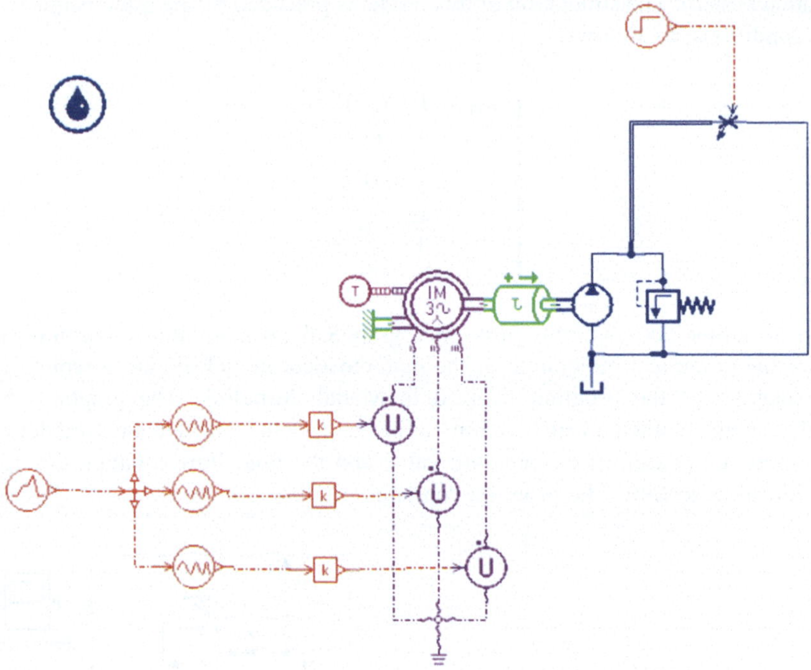

Fig. 4. Block diagram of the program realized in Amesim Simcenter

5 Results

The dynamic simulation of the system was performed in both simulation environments for a more accurate output of the mathematical model. A maximum simulation time was t = 5 [s]. At time t = 2 [s] a disturbance was artificially introduced, a change of the flow rate required by the actuating equipment (usually a hydraulic motor). The flow rate value decreases by 50% of the initial value, highlighting the dynamic behavior of the valve subjected to flow rate variations that may occur in the hydraulic system.

Simulation performed in Simulink environment uses the average flow rate relation (1), described above, and simulation performed in Amesim environment uses an ideal hydraulic pump. Figures 5a and b presents the difference between the two output flow rates of the simulated systems. Observing the oscillation pattern in Fig. 5a, a similar behavior of system's components is expected. The sampling period was chosen small enough (t_t = 0.0001 [s]) to highlight the continuous oscillation pump behavior.

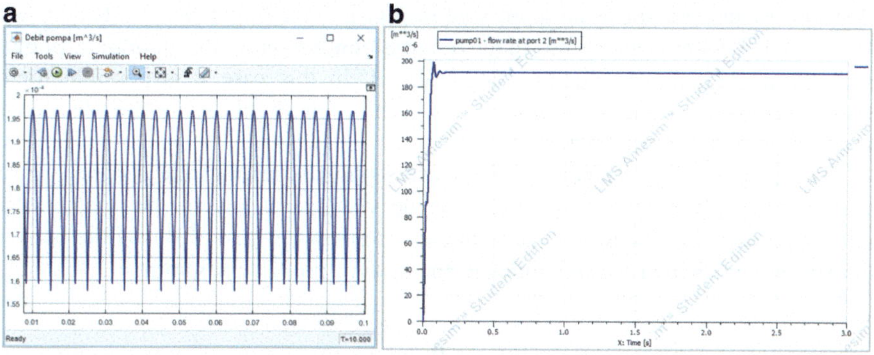

Fig. 5. a. Flow rate in Matlab. b. Flow rate in Amesim

On these two simulations, system pressure represents the variable of interest. Figure 6a represents the pressure obtained is Simulink environment and Fig. 6b represents the pressure obtained in Amesim environment.

It can be observed on both simulation environments an increase of the pressure on the initial time interval (approximately 0.15 s) until the signal stabilizes around the 30.93 bar (Fig. 6a), respectively 31.45 bar (Fig. 6b). The pressure is maintained at these values for 2 s, when the system flow disturbance occurs, corresponding to the 50% decrease of the hydraulic motor flow rate. In Simulink environment (Fig. 6a) there is a pressure increase of 6.2% and the signal stabilize at 32.85 bar. In Amesim environment (Fig. 6b) there is a smaller pressure increase (up to 31.84 bar) than in the Simulink environment due to the idealized components and conditions used for the system's model.

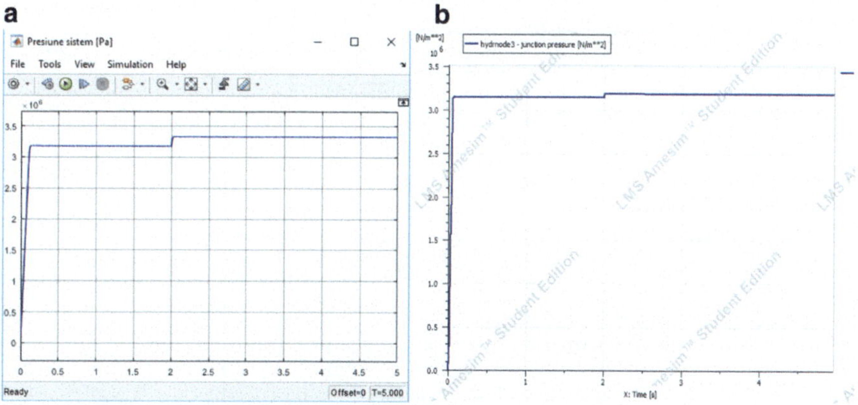

Fig. 6. a. System pressure in Matlab. b. System pressure in Amesim

Figures 7a presents the system pressure response to the flow rate disturbance, in the Simulink environment. First of all, the pressure oscillating pattern determined by the gear pump can be observed. At time t = 2 s, flow disturbance occurs. The transient time of the systems is about 0.3 s until it stabilizes. An increased stationary error can be observed.

Figures 7b presents the system pressure response to the flow rate disturbance, in the Amesim environment. Due to the ideal running conditions of the hydraulic pump, a linear pressure signal behavior can be observed. In this case it is possible for the system's pressure to stabilize extremely fast, which highlights again the ideal conditions of the system's components.

Valve opening (displacement of the valve's plug) is responsible for the pressure stabilization. Displacement of the plug can be visualized only on the mathematical model implemented in the Matlab Simulink environment. Figure 8 shows the behavior of the valve plug, which looks similar to the system pressure response.

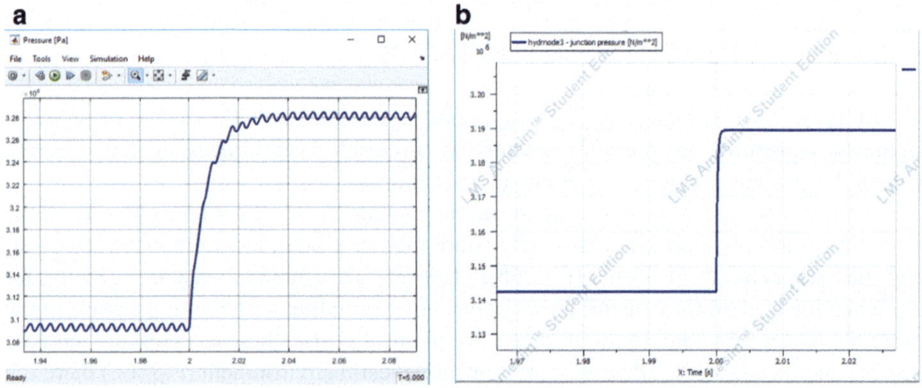

Fig. 7. a. Pressure signal response in Matlab. b. Pressure signal response in Amesim

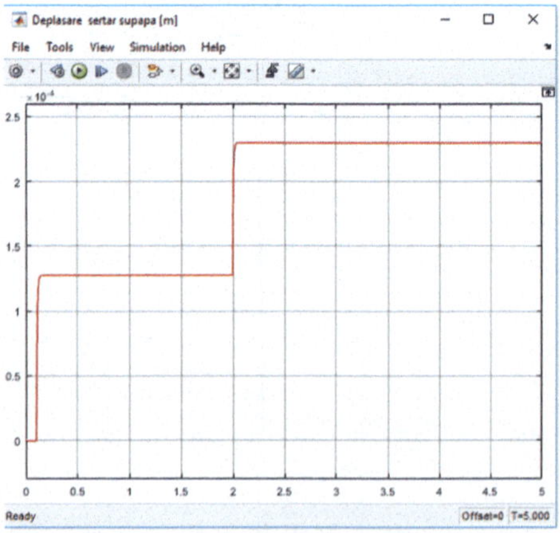

Fig. 8. Valve plug displacement

At the very beginning of the simulation, it can be observed how the valve remains closed until the system pressure reaches of the required value, Pr = 30 Bar. When the pressure in the system exceeds the desired value, the valve plug moves, changing the flow section of the valve; consequently, according to Eq. (3), the system pressure also changes. Thus, the valve adjusts the system pressure to the desired value.

6 Conclusions

The theoretical analysis of a hydraulic energy generation system concluded with the elaboration of mathematical model. This model was numerically integrated, using both the Matlab Simulink graphics programming environment and the Amesim Simcenter environment. The results obtained by the two methods are very similar, mostly because Amesim environment simulation runs using ideal components and conditions. Currently, the authors are preoccupied with the development of an experimental stand for analysis of a real-world hydraulic system. After that, the obtained results (from theoretical and experimental analysis) will be compared. In conclusion, the theoretical model presented can be used to approximate well enough the functioning of a hydraulic power generation system. The obtained model can be implemented into more complex drive systems.

References

1. Tataru, D.: Tehnica simularii. Aplicatii in biologie. Aspecte teoretice. Iasi: Stef (2010)
2. Broenink, J.F.: Introduction to Physical Systems Modelling with Bond Graphs, Modelarea Şi Simularea Sistemelor Mecatronice - 8 471 Univ. of Twente, Dep. EE, Enschede, Olanda (1999)
3. Modelul unui sistem. https://ro.wikipedia.org/wiki/Modelul_unui_sistem
4. Valer, D.: Proiectarea Sistemelor Mecatronice. Editura Politehnica, Timisoara (2007)
5. Roşca, R.: Elemente de mecanica fluidelor şi acţionări hidraulice, Editura "Ion Ionescu de la Brad", Iaşi (2015)
6. Landau, L., Lifchitz, E.: Physique théoretique. Mécanique des fluides", Ed. Mir Moscou (1989)
7. Savulescu, St. N., Dumitrescu, H., Georgescu, A., Bucur, M.: Cercetari matematice in teoria moderna a stratului limita, Ed. Acad. R.S.R. (1981)
8. Vasiliu, N., Vasiliu, D.: Acţionări Hidraulice Şi Pneumatice. Volumul I, Bucureşti (2004)
9. Avram, A.: Hidraulice Si Pneumatice. Editura Universitara, Bucharest (2005)
10. Jelali, M., Kroll, A.: Hydraulic Servo-Systems. Modelling, Identification and Control. Springer, London (2012)
11. Avram, M., Sârbu, V.-E., Spânu, A.-R., Bucşan, C.: Intelligent hydraulic power generating group. In: Proceedings of the International Conference of Mechatronics and Cyber-MixMechatronics (2018)
12. Avram, M., Nitu, C., Bogatu, L., Sarbu, V.: Theoretical analysis of an external gear pump-methods for determining the pumping capacity. Int. J. Mechatron. Appl. MechanicsOpen AccessVolume **2**(6), 182–190 (2019)

Author Index

G. I. Gheorghe (Ed.): ICOMECYME 2020, LNNS 143, pp. 243–244, 2020.
https://doi.org/10.1007/978-3-030-53973-3

The manufacturer's authorised representative in the EU is Springer Nature Customer Service Centre GmbH, Europaplatz 3, 69115 Heidelberg, Germany. If you have any concerns regarding our products, please contact ProductSafety@springernature.com

Printed and bound by CPI Group (UK) Ltd, Croydon, CR0 4YY
05/10/2026
02228122-0005